Environmental
Change
Network

This book is to be returned on
or before the date stamped below

The United Kingdom
Environmental Change Network:

Protocols for standard measurements
at terrestrial sites

Edited by
J M Sykes and A M J Lane

London: The Stationery Office

Centre for
Ecology &
Hydrology

Natural Environment Research Council

ISBN 0 11 702197 0

CONTENTS

ACKNOWLEDGEMENTS

The editors wish to thank the numerous individuals who have made contributions to the ECN programme or have been involved in the production of this publication but whose names do not appear in the text. In particular, we are indebted to the ECN Site Managers – John Adamson (Moor House), Roy Anderson (Hillsborough), John Bater (Rothamsted), Clive Bealey (Porton), Gordon Common (Sourhope), Stuart Corbett (Drayton), Paul Hargreaves (Rothamsted), David Henderson (Glensaugh & Sourhope), John Milne (Glensaugh & Sourhope), Mike Morecroft (Wytham), Alex Turner (Snowdon/Y Wyddfa), Deirdre Waddell (Alice Holt), Stuart Wright (Glensaugh) – who have provided most of the descriptive information on the sites contained in the Introduction, and have been responsible for testing the practicalities of implementing the Protocols. Our thanks are due also to those distinguished environmental and statistical academics and practitioners whose unstinting contributions to the preliminary and continuing work of the ECN technical committees have made the programme possible. Penny Ward has had the daunting task of finalising the preparation of the manuscript for publication, and we are grateful to her for her skill and forbearance.

PREFACE

ECN in prospect and in retrospect

Monitoring was an unfashionable word in the 1980s. It was associated with routine, unscientific observations which contributed little to the understanding of environmental change. It was not linked to research designed to determine causal relationships and to predict future conditions. Whether or not this perception of monitoring was true, the position has changed. Recognition of the lack of long-term, rigorous, comprehensive and quantitative information on the state of our environment has come to the fore, precipitated by concerns such as biodiversity, widespread pollution, and predictions of climate change. The critical observations of stratospheric ozone and atmospheric carbon dioxide levels, obtained through the persistence of individual scientists, have added weight to the arguments for sustained, long-term observations.

The need for long-term observations was a major motivating force behind the formation of the UK Environmental Change Network (ECN). Another factor was the recognition that many different organisations, for very different reasons, have well-established field stations or sites on which long-term observations and experiments were being made. With adaptation and co-ordination of the observations to obtain comparability, co-operation between the organisations could provide a network of monitoring sites across the UK, sampling most of the main land uses and climatic regions. Thus, a national network could be established, with relatively little effort and cost, by building on existing facilities. There was the added advantage of close association with some of the finest research experience in the country.

A third factor, which was part of ECN thinking from the start, was the concept of integrated monitoring. There are many projects which are concerned with detecting long-term change in particular aspects of the environment (eg water quality, acid rain, or ozone) or with particular groups of organisms (eg moths, butterflies, birds, or plants), or with particular processes (eg crop production or tree health). However, these represent only parts of the environment and ecosystems. Integrated monitoring is the measurement of related variables in different biotic and abiotic compartments and co-ordinated in space and time to provide a comprehensive picture of the system under study. As argued by Munn (1988), 'When based on an interconnected picture of the environment and the biosphere (through the notion of biogeochemical cycling of trace substances, for example), the monitoring system is much more likely to be responsive to detecting surprises than if it consisted of several disconnected components (an air monitoring network, a water quality network, etc)'. Such integrated systems have been operated or planned for various sites and potential networks (eg Santolucito 1991), but the only comprehensive network in which the approach has been put into effect is

1

the International Co-operative Programme on Integrated Monitoring of Air Pollution Effects on Ecosystems (IMP), operating within the UN-ECE Convention on Long-Range Transboundary Air Pollution (UN-ECE 1993) and focused on forested catchments.

The general principles on which ECN was established were reasonably easy to define. Putting them into practice was another matter. However, the initial concentration on terrestrial sites was successful in getting agreement between nine UK Government departments and agencies to support the co-ordinated measurements at their individual sites, and to contribute the data to a central database funded by the Natural Environment Research Council. This was a major achievement which has been followed by expansion, particularly in freshwaters, so that the total number of sites is now 49, with 14 collaborating organisations.

The establishment of the network of sites is one step; defining the variables and methods to cover such a broad range of components and systems is more difficult. This required considerable effort by the ECN Technical Working Group, in consultation with many scientists. It would have been ideal to use past data, combined with scenarios or models, to predict the response of systems to environmental change, and then to select the variables which would provide most information. Such an approach was not feasible, given the state of predictive science and the need to maintain momentum. A degree of pragmatism was required and the selection of variables was based on expert knowledge and practicality, but there is a clear logic in the selection of interconnected driving, state and response variables which will allow testing of relationships. Physical and chemical variables were much easier to select than were biotic variables because the options were more limited and the experience greater (cost became a key criterion). The options for biological variables or indicators were much greater, especially with the wide variety of sites and the requirement for comparability and feasibility, a problem which is faced in other monitoring networks. These problems were addressed by drawing on the experience of other monitoring programmes and a breadth of ecological knowledge. The resulting selection of biotic variables for the terrestrial sites includes a reasonably balanced combination of animal species which represent functional groups (herbivores, insectivores, decomposers), as well as wider taxonomic groups (Lepidoptera, Chiroptera, Aves), and the more obvious plant species composition. Atmospheric soil and water chemistry provide information on biogeochemistry.

The measurements are now in progress, a comprehensive database is beginning to accumulate, and quality controls are working well. The database is at the heart of ECN. It is designed and operated to ensure that all the data collected across the network are properly validated, maintained, documented and archived for posterity. But the data are not merely stored – they are made immediately available for short-term uses in a variety of areas. The dissemination of data and information through research papers, annual data digests, national state-of-the-environment

2

reports and even direct database access across the Internet are just some of the ways in which ECN is developing the use of its data.

The publication of the Protocols is a step designed to provide detailed, citable background for anyone using data from ECN. It also increases the opportunity for wider debate on the technology of integrated monitoring; it is part of the process of quality assurance, and it may also assist other networks in their planning.

ECN has come a long way since the germination of ideas began in 1986, and its formal establishment at the official launch in 1992. It is now in the early stages of succession. What of the future? Some activities which are likely to develop are:
- a period of stability to allow accumulation of sufficient data to test the value of individual variables;
- testing of a few other variables (eg biomass, plant chemistry, soil biota, decomposition) for possible addition to the suite of observations, although options for expansion are limited by support costs;
- a small and gradual increase in the number of terrestrial sites to improve the network's environmental and land use cover;
- increasing interaction with sectoral monitoring projects which can allow wider geographical assessment of changes in individual components;
- further analysis of past data from individual sites to enhance the time series, define variability, and explore responses to environmental variation;
- enhancement of the links between the terrestrial and freshwater components of ECN;
- development of a 'rapid response procedure' to capitalise on the existence of the network in the event of extreme environmental events;
- focusing of results from national research programmes in order to clarify, and possibly test, the distribution of changes that are likely to occur across the network.

The individual sites are crucial pieces of the ECN jigsaw. It is the strength of the site researchers on which the network depends. But it is the network as a whole which is the instrument for the detection of change. One of the most important capacities is exploring the extent to which changes detected at one site are observed at others: are the changes national, regional or merely local? It is here that the links with other networks and research programmes have great potential.

Thus, ECN is only part of the system to detect and understand environmental change and its consequences in the UK. It represents the intensive, fine-scale level of resolution at which variables are measured continuously at short time intervals . This identifies a limitation of ECN, namely the small number and non-random distribution of sites, a feature which limits the rigorous extrapolation of results. But ECN does not function in isolation. It is linked to the various sectoral observing systems and, in particular, to the national network of the Countryside Survey which represents a larger, more extensive, stratified sampling system with observations made at five to ten year intervals (Barr *et al.* 1993). Further, the Land Cover Map (Fuller, Groom & Jones 1994) provides complete cover of Great Britain obtained from satellite imagery, with the option of frequent observation. The combination of these three levels of resolution provides Britain with a strategic monitoring design which is one of the most complete, and probably unique, in the world. At a time when there are major efforts by international organisations to develop systems to detect changes over large areas, such as the European Environment Agency and the Global Terrestrial Observing System (GTOS), the experience of the UK in establishing ECN and related systems has much to offer the wider environmental community.

Finally, it is a pleasure to acknowledge the contribution of so many individuals and organisations to the establishment of ECN in general, and to the publication of this handbook of Protocols in particular. Especial thanks go to members of the Technical Working Group (Chairman, Professor Mike Hornung), the Statistics and Data Handling Working Group (Chairman, Professor Richard Cormack), and their successor the Statistical and Technical Advisory Group, for their persistence in a long and difficult exercise; to Mike Sykes, the first ECN Co-ordinator, Terry Parr, his successor, and members of the Central Co-ordination Unit; to the unsung Site Managers whose testing and experience in implementing the Protocols has been essential; and to the departments and agencies without whose visionary commitment ECN would not have been possible.

Professor O.W. Heal
Chairman, ECN Steering Committee

References

Barr, C.J., Bunce, R.G.H., Clarke, R.T., Fuller, R.M., Furse, M.T., Gillespie, M.K., Groom, G.B., Hallam, C.J., Hornung, M., Howard, D.C. & Ness, M.J. 1993. *Countryside Survey 1990: main report.* (Countryside Survey 1990 vol.2.) London: Department of the Environment.

Fuller, R.M., Groom, G.B. & Jones, A.R. 1994. The Land Cover Map of Great Britain: an automated classification of Landsat Thematic Mapper data. *Photogrammetric Engineering & Remote Sensing,* **60**, 553–562.

Munn, R.E. 1988. The design of integrated monitoring systems to provide early indications of environmental/ecological changes. *Environmental Monitoring and Assessment,* **11**, 203–217.

Santolucito, J.A., ed. 1991. Long-term ecological monitoring for global change. *Environmental Monitoring and Assessment, Special Issue,* **17**, 1–78.

UN-ECE. 1993. *Manual for integrated monitoring. Programme Phase 1993–1996.* (Environmental Report 5.) Helsinki: Environmental Data Centre, National Board of Waters and the Environment.

Chapter 1   INTRODUCTION

Monitoring global environmental changes

The occurrence of far-reaching changes in the earth's environment is now well recognised by scientists, politicians and public, and this recognition has generated a growing national and international interest in detecting and monitoring environmental changes. At the global scale these changes are caused by largely man-induced changes in climate, atmospheric composition and land use (International Geosphere–Biosphere Programme 1992), factors which also operate at regional and national scales where they are often exacerbated, or occasionally mitigated, by local factors.

International efforts to obtain reliable information on the responses of natural and managed ecosystems to global environmental changes have burgeoned during the last decade and have resulted in the involvement of several organisations and their associated actual or planned networks. An overview of existing organisations and networks is provided in Heal, Menaut and Steffen (1993) and a comprehensive listing is provided by Fritz (1991).

Environmental monitoring in the United Kingdom

The UK has a long history of environmental monitoring, sampling being carried out by a multiplicity of organisations for a wide variety of purposes. The majority of this monitoring is designed to ensure that there is compliance with policies of environmental regulation set out in international, national and local agreements. In addition, monitoring provides information on the effectiveness of policies already being implemented and may lead to proposals for new or modified policies or actions, especially where early warnings of environmental changes have been recognised. Finally monitoring is concerned with the measurement of background levels and the provision of benchmark data for research and policy purposes, as well as with reassuring the public. McCormack (1990) has provided an overview of UK environmental monitoring systems, especially in relation to the concepts, methods and strategies used.

Most UK environmental monitoring is de-centralised, the majority of sampling being carried out by local government, other public sector bodies including inspectorates, individual factories and at both central Government and other research laboratories. Most monitoring is also sectoral, in the sense that particular monitoring systems have been devised and developed to relate to a particular sector of the environment. Various central UK Government departments have responsibilities in the areas of, for example, air quality, water, land, soil, natural resources, flora and fauna. There is co-ordination of monitoring programmes within sectors, eg air monitoring networks, and also across sectors where this approach is necessary, eg monitoring radioactivity in air, drinking water, the sea, and in agricultural products.

Non-governmental organisations play an important role in environmental monitoring; they often optimise the involvement of the large pool of available amateur expertise and inform the public of environmental changes which are taking place.

There is a growing movement towards harmonised monitoring which, whilst it may be carried out by different organisations, produces reliable data, capable of comparison and integration at both national and international levels. The national environmental data resource is largely concentrated in databases and geographical information systems at designated Environmental Data Centres. Contributions are made from UK environmental monitoring programmes to international programmes such as the United Nations Economic Commission for Europe (UN-ECE) International Co-operative Programme on Integrated Monitoring of Air Pollution Effects on Ecosystems (IMP), and the United Nations Environment Programme (UNEP), Global Environmental Monitoring System (GEMS) and Global Resources Information Database (GRID).

Further co-ordination of environmental monitoring is being achieved by the use of an area-based approach, using a system of land classification developed by the Institute of Terrestrial Ecology. The classification, which allows 1 km squares in Great Britain to be allocated to 32 land classes, provides an agreed objective sampling and stratification framework for national ecological surveys and monitoring (Bunce *et al.* 1996)

The Environmental Change Network (ECN)

The need for a general-purpose network designed for long-term, integrated environmental monitoring in the UK, especially in relation to current or future major anthropogenically induced factors, has been emphasised on numerous occasions and has eventually led to the formation of ECN (Tinker 1994). As early as 1976 a Natural Environment Research Council (NERC) Working Party on Biological Surveillance (NERC 1976) had noted the need for detailed surveillance at a limited number of sites with the objective of observing natural changes on a year-to-year and long-term basis. It was suggested that protected areas and sites with intensive research programmes or where substantial information was already available should be used for this purpose. Almost a decade later a House of Lords Select Committee on Science and Technology (1984) recommended that the effects of agricultural practices should be monitored by 'a small and highly selective network of projects . . . . to give early warning of environmental consequences'.

As a consequence of these recommendations and a wide recognition in the scientific community of the need for a network which would meet the requirements of environmental change research and monitoring, NERC undertook, in 1986, to consult UK Government departments and research organisations primarily concerned with agriculture and the environment to explore the setting up of such a network. A working group on long-term reference sites was set up which produced a series of recommendations for the establishment of a national network of sites which would meet the requirements of different interested organisations (NERC 1986). These recommendations were taken up and put into effect by a consortium of agencies which agreed to contribute to the operation of the network, which became the UK Environmental Change Network (ECN).

The rationale for proposing a network of sites is summarised by Heal (1991), as follows.

1. Study sites are an essential component of ecological research. To answer questions on changes in the environment, we need sites which represent the main environmental, ecological and management variations in the UK. While studies of some individual topics will require other sites with particular characteristics, a 'core' network will provide opportunity to use existing information on the related topics.

2. Long-term studies are required to monitor changes external to the system which take place gradually or at infrequent intervals. Responses to those changes may occur through species or processes which have a slow turnover time or through a series of linked short-term events, the results of which are only apparent in a long-term study.

3. In addition to delayed and serial responses, it is also necessary to distinguish between the different factors which cause, or interact to cause, change. For these reasons it is important to have sites with integrated or multi-media monitoring and to carry out both observational and experimental research.

4. The scientific case for a network of long-term study sites in the UK is strong. Information on environmental changes and on their consequences is a serious need in UK Government. By concentrating on established sites, the cost of creating such a network can be kept to a minimum.

ECN objectives

The objectives of ECN are as follows.

*   To obtain uniform and comparable long-term datasets at selected sites by means of measurement at regular intervals of variables identified as being of major environmental importance.

*   To provide for the integration and analysis of these datasets so as to identify environmental changes and to improve understanding of the causes of such changes.

- To make these long-term datasets available as a basis for research and for the prediction of possible future changes.

- To provide, for research purposes, a range of representative sites where there is good instrumentation and reliable environmental information.

ECN design

The ECN aims to monitor changes in selected biota in addition to the physical and chemical environment. The programme thus falls within the definition of 'ecological monitoring' (Hinds 1984). It is not surprising, therefore, that the design of the ECN has encountered the problems which the author identifies as needing to be overcome in successful ecological monitoring designs and which can be summarised as:
- selecting and quantifying specific entities within the continuous spatial and temporal flux;
- specifying appropriate replication standards in a world that is full of unique places;
- expense.

The need for long-term observations in ecology has been set out by Likens (1983) and Strayer *et al.* (1986) and summarised by Woiwod (1991), who also discusses the scientific, political and personal problems associated with long-term experiments and observations. The problems of sustaining a long-term programme such as ECN are exacerbated to some extent by the participation of many organisations, all of which have different objectives, are publicly funded, and are unable to commit funds for more than three to five years in advance. Nevertheless, it was believed that the programme could be sustained if it was not too ambitious, had a well-defined concept and organisation, was able to operate successfully within agreed target budgets, and if the network as a whole provided added value to the individual contributions of sponsoring agencies. The initial steps to be taken were as follows.

1. Select a series of variables related to climate, pollution and land use, changes in which would drive the states of a second set of 'response' variables. Both driving and response variables should be interpretable, informative, comparable and repeatable between sites and times, and response variables should be sensitive to changes in the driving variables. Within these constraints they should also, where possible, be simple and cheap and avoid labour-intensive operations. The variables should be selected so as to be measurable at each of a series of sites which may have a wide range of climatic, topographic, soil and vegetation (crop) conditions.

2. Establish agreed, strict and clear protocols for the sampling and recording system to be used for measuring each variable, for chemical analysis where necessary, and for quality control and assurance of the data.

3. Establish methods for managing and storing the data.

An informal Workshop was held in September 1990 at which researchers, academics and environmental practitioners discussed organisation, possible sites, and the rationale and technicalities of a range of possible variables to be included in the programme. Subsequently the ECN Steering Committee set up two Working Groups to develop further the ideas and views from the Workshop.

A **Technical Working Group**, comprised of experts in areas of environmental science relating to the programme and chaired by Prof M Hornung, refined the list of suggested variables and formalised the methods of measurement to be used. Variables were selected to meet, as far as possible, the criteria set out in (1) above and resulted in a set of 'core measurements' which were to be undertaken at all of the ECN sites. A second set of variables, 'priority additional measurements', were considered as being desirable but not essential and therefore only to be undertaken if funds allowed. In practice, funds have been, and are likely to remain insufficient to allow such additional variables to be measured and these are not discussed further in this account. Where national sectoral monitoring schemes were already operating, their sampling design and methods were adopted by ECN. This provided added value for ECN in that data from its sites could be analysed in a broader regional and national context, whilst ECN could provide a comprehensive set of relevant environmental data for the sectoral networks. Examples of such sectoral networks are the Butterfly Monitoring Scheme, Rothamsted Insect Survey, Common Birds Census, Breeding Bird Survey and UK Precipitation Composition Network.

A **Statistics and Data Handling Working Group**, comprised of statisticians and computer scientists and chaired by Prof R M Cormack, advised on all matters relating to the statistical design of the programme, and in particular on the spatial location of sites and observation points within sites, the selection of variables and the methods and temporal frequency of measurement. The Group also advised on methods for storing and managing data, methods for analysing new and old data, and on the resources needed to meet their recommendations. Cormack (1994) discusses some of the statistical aspects of planning the network and the main questions put to the Group, together with the answers given. Some of the considerations are described below.

•   For most environmental variables, sampling variation is likely to be much greater than analytical variability; however, the necessary degree of replication is difficult to define because information on distributional form or variance components of the selected variables is often unavailable.

•   The intensity of destructive sampling in certain areas and the consequent need to avoid the possibility of one sample interfering with another affect the spatial layout of sample points.

- Objectivity in taking a sample may be more fundamental to the aims of the programme than the choice between random and systematic samples.

Protocols for data analysis have not been set down but the Group formulated some general principles.

- Analytical procedures to be applied routinely to data as they are collected should be simple but not make simplistic assumptions; more detailed parametric methods should be adopted to test specific hypotheses as they arise.

- Cusum charts and graphical procedures of exploratory data analysis should be widely used.

- A modelling paradigm should not be embraced too early: data dredging by techniques such as multiple regression should not be used without comprehensive cross-validation, and time-series modelling should be carried out only after the formulation of specific hypotheses.

- Space/time analyses may be informative as records accumulate but assumptions of spatial stationarity may need careful study.

Reciprocal feedback between the two Working Groups was an important part of the development process.

Development of the network

Although ECN was conceived of as a programme covering a wide range of natural, semi-natural and managed ecosystems, the need for early implementation of the programme led to the adoption of a step-by-step approach to network establishment. It was decided that attention would first be focused on setting up a network of terrestrial sites, to be followed as soon as possible by a parallel and linked network of freshwater sites, to include rivers and lakes. Each contributing agency agreed to provide one or more sites and the resources to carry out an agreed suite of ECN measurements, or to provide equivalent resources to support the general operation of the network. A list of current contributors is provided in Table 1, which includes agencies responsible for both the terrestrial and freshwater sites, though information on the latter is not provided in this publication but will appear in a later volume.

Site selection

A preliminary list of 140 possible sites was compiled, based on their being currently or recently active research sites which could provide historic data on some aspects of environmental research. The following criteria were then used to refine the list and to select a series of 24 target sites which, it was believed, could constitute a terrestrial network for the United Kingdom:

Table 1. Supporting agencies and sites in ECN

Agency	Sites/support	Site type
Biotechnology & Biological Sciences Research Council	•North Wyke Research Station	Lowland grassland
	•Rothamsted Experimental Station	Arable
Countryside Council for Wales (jointly with Welsh Office)	•Y Wyddfa/ Snowdon NNR	Upland grassland
Department of Agriculture for Northern Ireland	•Agricultural Research Institute, Hillsborough	Lowland grassland
	•3 freshwater sites	River & lake sites
Department of the Environment	•Support for Central Co-ordination Unit	
	•4 freshwater sites	River & lake sites
Department of the Environment for Northern Ireland	•2 freshwater sites	River sites
English Nature	•Site & facilities at Moor House –Upper Teesdale NNRs	
Environment Agency	•13 freshwater sites	River & lake sites
Forestry Commission	•Alice Holt Forest	Woodland
Ministry of Defence	•Porton Down	Chalk grassland
Ministry of Agriculture, Fisheries & Food	•ADAS Drayton	Mixed farming
	•Soil Survey & monitoring at English & Welsh sites	
Natural Environment Research Council	•ECN Central Co-ordination Unit	
	•Moor House–Upper Teesdale NNRs	Upland grassland & blanket bog
	•Wytham (site provided by Oxford University)	Woodland & arable
	•4 freshwater sites	River & lake sites
Scottish Environment Protection Agency	•12 freshwater sites	River & lake sites
Scottish Office, Agriculture, Environment & Fisheries Dept	•Glensaugh Research Station	Upland grassland
	•Sourhope Research Station	Upland grassland
Welsh Office (jointly with Countryside Council for Wales)	•Y Wyddfa/ Snowdon NNR	Upland grassland

- good geographical distribution covering a wide range of environmental conditions and the principal natural and managed ecosystems;
- some guarantee of long-term physical and financial security;
- known history of consistent management;

- reliable and accessible records of past data, preferably for ten or more years;
- sufficient size and opportunity to allow further experiments and observations.

The target sites met most of these criteria but it was inevitable that, in a network for which there was very limited central funding, the second became the primary criterion. Whilst physical security was, for the most part, assured by the sites being located at or being associated with existing research stations, assurances of financial security for what was envisaged as a long-term project with an initial life expectation of at least 30 years were difficult to obtain.

The terrestrial network was thus founded, in January 1992, with eight of the target sites (Drayton, Glensaugh, Hillsborough, Moor House–Upper Teesdale, North Wyke, Rothamsted, Sourhope, and Wytham) being committed and with the expectation that other sites would join later. By 1996 the network had expanded to a total of 11 sites by the addition of Alice Holt, Porton Down and Snowdon/Y Wyddfa. None of the additional sites was in the original list of target sites but each increased considerably the geographical distribution and representativeness of the network.

It is intended that this 'spine' of representative sites should be extended in the future and supplemented by the recruitment of sites which will provide replication within the major axes of UK environmental variation.

Network sites

Figure 1 shows the locations of the 11 terrestrial sites as well as the 38 lake and river sites of the freshwater network which started operating in 1995. The terrestrial network sites are as follows.

Alice Holt Forest, Surrey, England (Lat 0° 50'W; Long 51° 10'N)
Sponsor: Forestry Commission

Alice Holt Forest lies in the Weald, between the North and South Downs in southern England. The Forest is situated on a very gently sloping plateau at an altitude of 110–125 m and has a mean annual rainfall of 768 mm (1956–94). The soils range from brown soils and podzols to surface-water and ground-water gleys. The Forest, which has belonged to the Crown since the time of William I who was responsible for the creation of the Royal Forest as a hunting preserve, was subjected to severe felling for warship building programmes during the 17th and 18th centuries. Replanting of oak (*Quercus robur*) took place subsequently and some of the plantings between 1815 and 1820 still survive. By 1881 it had become clear that oak on the leached and less fertile soils was not growing satisfactorily and some of this was cleared and the ground replanted with conifers. The forested area of 850 ha is mainly of Corsican pine (*Pinus nigra* var. *maritima*) but 140 ha of the oak planted in 1820 remain, and there are 21 ha of 'unproductive' forest. Vegetation is diverse, ranging

12

Freshwater sites

▲ River sites   ☐ Lake sites

1	Eden (Cumbria)	A	Upton Broad
2	Esk	B	Hickling Broad
3	Coquet	C	Wroxham Broad
4	Exe	D	Windermere
5	Wye	E	Esthwaite Tarn
6	Lathkill	F	Loch Leven
7	Cringle Beck	G	Scoat Tarn
8	Frome	H	Llyn Llagi
9	Bradgate Brook	I	Lochnagar
10	Bure	J	Loch Lomond
11	Old Lodge		- Cailness
12	Stinchar		- Creinch
13	Lower Clyde	K	Loch Katrine
14	Allt a'Mharcaidh	L	Loch Davan
15	Spey (Fochabers)	M	Loch Kinord
16	Tweed (Galafoot)	N	Loch Dee
17	Eden (Fife)	O	Lough Neagh
18	Cree	P	Lough Erne
19	Faughan		
20	Garvary		
21	Bush		
22	Trout Beck (Moor House)		

⬤ Terrestrial sites

Figure 1. Location of ECN freshwater and terrestrial sites

13

from base-rich communities of the ash (*Fraxinus excelsior*), field maple (*Acer campestre*), dog's mercury (*Mercurialis perennis*) type of woodland, base-poor communities of oak, bracken (*Pteridium aquilinum*), and bramble (*Rubus fruticosus*) on the sand/gravel soils, and oak and birch (*Betula pendula*) on the heathy, acidic soils. These acidic areas are generally planted with either Scots pine (*Pinus sylvestris*) or Corsican pine and there is often abundant natural regeneration of birch.

The Forestry Commission has had a research establishment at Alice Holt Lodge, formerly used as a private residence, for almost 50 years, and climatic data have been recorded at the site since the late 1940s. There is other monitoring on the site which is particularly relevant to ECN, such as the Rothamsted light trap which has been used continuously to collect macro-lepidoptera since May 1966; a more recent long-term project, the Forest Condition Survey, began in 1984 and the Forest is now a European Union site for monitoring the effects of atmospheric pollution on trees. A variety of other experimental work in Alice Holt Forest has been carried out on forest insects, pathogens, silviculture, mensuration and ecology. There is an active current research programme undertaken by the Forest Authority's Research Division from Alice Holt Lodge, which includes research on forest soils, hydrology, atmospheric deposition, insect ecology and conservation, pest management, management of habitats and landscapes, including deer, squirrels, bats, raptors, seed testing, nursery techniques, pathology, yield and carbon storage modelling. The site also has permanent sample plots, an arboretum, registered seed orchards, and clone banks, and a farm woodland demonstration plot.

Drayton, Warwickshire, England (Lat 1° 45'W; Long 52° 12'N)
Sponsor: Ministry of Agriculture, Fisheries and Food

Situated 5 km west of Stratford-upon-Avon in the English Midlands, Drayton is part of a Research Centre, one of a network of five such Centres operated by the Agricultural Development and Advisory Service, a central UK Government agency which provides consultancy and advice services to the land-based industries in England and Wales. The farm at Drayton occupies 190 ha and is underlain by Lower Lias rocks comprising calcareous clays inter-bedded with thin, brashy, fossiliferous limestone. The depth of the overlying clay drift varies from a few centimetres to more than 1 m. The soils are very heavy, with clay contents of 50–75%. They have poor natural drainage, are extremely difficult to work and compaction and smearing can occur easily if they are cultivated when wet; however, they weather readily, are generally alkaline and have a high potash level, though available soil phosphate tends to be low. The altitude of the farm is 40–80 m, with gentle gradients falling towards the south and east. The climate is typical of the central lowlands of England – fairly dry, with cold winters and warm summers. Average rainfall (1941–93) is 620 mm. February (39 mm) is the driest month and August (68 mm) is wettest. Average minimum air temperature in January–March is <2°C whilst average maximum temperature in the period June–September is >19°C.

Drayton is representative of the mixed farming systems used on the Midland clays. The landscape comprises a patchwork of grass and arable fields with an average size of only 4.9 ha, enclosed by hedgerows which are dominated by hawthorn (*Crataegus monogyna*) and which have a total length of 17 km on the farm. There are few mature trees, partly due to the effects of Dutch elm disease, and no areas of established woodland. Watercourses include two small permanent streams, drainage ditches and two ponds.

The farm has the following cropping regimes:
- intensive, continuous, arable cropping of autumn-sown, combinable crops based on wheat, with field beans and oilseed rape as break crops;
- rye-grass leys of four to six years' duration, alternating with two or three years of arable (wheat) crops;
- permanent or long-term rye-grass leys.

In addition, 13 ha of formerly arable land are currently occupied by farm woodland. Livestock on the farm includes a flock of 300 ewes which are kept for lamb production, a flock of 140 wethers, and dairy-bred calves.

Experimental work has been carried out at Drayton since 1940. Recent research has included studies of various aspects of weed, pest and disease control on arable crops, including determination of pesticide efficacy, development of forecasting systems for pest and disease control, and comparisons between chemical and cultural control of weeds. Two long-term studies are measuring the economic, agronomic and environmental effects of reducing inputs of pesticides and nitrogen compounds into farming systems. Other projects aim to quantify the effects of pesticides on non-target organisms. Recent developments in setting aside formerly arable land have led to monitoring of vegetation development, pests, diseases and soil nitrogen during rotational and non-rotational set-aside regimes and to the evaluation of farm woodland establishment and management techniques. Livestock research includes forage production and evaluation and animal health studies, whilst a related study is examining potential pollutants in runoff from animal wastes applied to grassland.

Glensaugh, Grampian Region, Scotland (Lat 2° 30'W; Long 56° 51'N)
Sponsor: Scottish Office, Agriculture, Environment & Fisheries Department

The Macaulay Land Use Research Institute's Glensaugh Research Station is situated on the south-eastern edge of the Grampian Mountains, 55 km south-west of Aberdeen. It extends to 1125 ha, of which 970 ha are semi-natural vegetation, 150 ha short-term and permanent grassland and 5 ha deciduous and coniferous woodland. There is also a small loch. The land rises from an altitude of 100–450 m. Mean annual rainfall at 195 m is 1040 mm and annual mean daily sunshine hours are 3.83. At higher altitudes there can be considerable snow cover for periods in the winter.

The site is bisected on its east/west axis by the Highland Boundary Fault. North of the Fault soils have been developed on drifts derived from a variety of Upper Dalradian schists, whilst south of the Fault soils have been developed on till derived from an admixture of Lower Red Sandstone sandstone and Dalradian schists. The whole area has been extensively glaciated so that the hill tops are well rounded, and deep glacial till deposit covers the lower slopes with thinner drift on the upper slopes on both sides of the Fault. North of the Fault brown forest soils and podzols are most common on the lower slopes, whilst peaty podzols cover the higher slopes and there are extensive areas of peat with an average depth of 2 m on the summit areas. South of the Fault humus-iron podzols predominate on the lower slopes, changing to shallower, stony peaty podzols at higher elevations. A number of streams drain the small catchments related to the four hills on the site, running south into the River North Esk, some of them *via* Loch Saugh.

Dry heather (*Calluna vulgaris*) moor is extensive on the more freely drained slopes, the associated species being mainly blaeberry (*Vaccinium myrtillus*) and wavy hair-grass (*Deschampsia flexuosa*). The dry heather communities gradually merge into heather/cottongrass (*Eriophorum vaginatum, E. angustifolium*) blanket bog on the deep peat of the hill tops north of the Research Station. A regular burning rotation of the moor is aimed for, but unsuitable weather over the last five years and the proximity of large forest plantations to leeward of the prevailing wind have prevented burning at the desired frequency.

Rye-grass/clover (*Lolium perenne/Trifolium* spp.) swards are found at most lower altitudes, half the area being permanent pasture and the other half in five-year leys. About half of this grassland is cut for silage each year. On the lower hill slopes are both species-rich and species-poor bent/fescue (*Agrostis/Festuca*) communities. Bracken is associated with these communities, particularly on south-facing slopes and at the interfaces of bent/fescue and heather communities.

The principal agricultural enterprise is the production of weaned and finished lambs from a flock of 100 ewes, but in addition there are 60 beef cows and 120 farmed red deer (*Cervus elaphus*) hinds which, like the ewes, are grazed on both the sown grassland and semi-natural vegetation.

Glensaugh has been used as a research facility for over 50 years and in consequence there is a good record of past resource management, as well as meteorological data over the same period and many data from long-term experiments and monitoring, some of it relevant to ECN. The Station provides land resources for a wide range of research including agroforestry, vegetation dynamics of heather moorland, ruminant foraging behaviour, biology of red deer, acidification of catchments and the dynamics of nutrients and pollutants in the soil. Much of the research is collaborative with other research institutes and universities.

Hillsborough, Co Down, Northern Ireland (Lat 6° 05'W; Long 54° 27'N)
Sponsor: Department of Agriculture for Northern Ireland

The Agricultural Research Institute, Hillsborough, lies 12 miles south-west of Belfast and was set up in 1927 in Large Park, then part of the Hillsborough Castle Estate. The Institute controls 200 ha of Large Park and the land is mostly in grass, interspersed with 200 ha of State Forest. Hillsborough Lake occupies 15 ha on the interface between Large Park and the town of Hillsborough and is surrounded by amenity woodland which is open to the public. The land rises from an altitude of about 110 m around the town to 170 m at its southern boundary. The underlying rock is Silurian slate overlain by varying depths of glacial clay till and with alluvial deposits between the drumlins in the northern part of the site, where shallow peats also occur locally. The climate is oceanic, ie mild and wet, and data for the period 1987–91 show a mean maximum daily temperature of 12.2°C and a mean minimum daily temperature of 5.9°C. Total annual precipitation for this period ranged between 737 mm and 973 mm and mean daily hours of sunshine were 3.87. In winter, while prolonged periods of lying snow are unusual, the site is exposed to cold winds from the north and west, and summers are also often cool and cloudy; these conditions are typical of the north-west of the British Isles. Lack of insolation combines with relatively heavy, water-logged soils to produce humid, cool, ground conditions which suit relatively cold-adapted, hygrophilous organisms and there is a year-round problem with slug populations on horticultural and other crops, as well as considerable local damage to earthworm populations due to predation by the introduced New Zealand flatworm (*Artioposthia triangulata*).

The Institute provides resources for research, primarily into beef and dairy cattle nutrition and the management of grassland, but a wide range of other, often collaborative, research activities are undertaken to help improve the competitive position of the Northern Ireland agriculture and food industries. Some current research areas are:
• factors affecting carcase composition in beef cattle;
• bull beef production on forage diets;
• continental beef production;
• development of near infra-red reflectance spectroscopy for the prediction of forage metabolisable energy content and the intake of forages by dairy and beef cattle;
• effects of silage fermentation type and the nature of concentrates fed on milk production in dairy cattle;
• sheep production from grass and sheep response to silage/ concentrate feeding;
• quantitative studies of the nitrogen cycle in grazed grassland;
• reduced input systems for winter barley, oilseed rape and potato crops;
• flax preservation and retting;
• the fertilizer value of sewage sludge;
• investigation of methods to improve the resistance of concrete to attack by silage effluent.

Conditions are good for grass growth; perennial rye-grass or mixed rye-grass/clover swards in Northern Ireland produce in excess of 30 t ha^{-1} yr^{-1} of silage. All but a few hectares of the site are sown with perennial rye-grass and re-seeded on a seven-year cycle. The bulk of this is harvested in a three-cut silage system or grazed by sheep or dairy cattle. Within this regime are small areas of permanent pasture, including the ECN Target Sampling Site (TSS). The TSS is managed on a two-cut silage system with the sward grazed early and late in the season by sheep. This is regarded as more typical of best local practice in the north Co Down area than the more intensive management adopted elsewhere on the farm.

The State Forest in Large Park is mostly coniferous plantations, with Norway spruce (*Picea abies*) and hybrid larch (*Larix* x *eurolepis*) as the main species. There is a sizeable area of mixed amenity woodland around Hillsborough Lake but elsewhere only scattered blocks of pure broadleaved trees with beech (*Fagus sylvatica*), sycamore (*Acer pseudoplatanus*) and oak predominating.

There is a long history of research at the site which has had a manual meteorological station for over 40 years and, more recently, automatic stations at some of the experimental sites. There are also precipitation collectors at several locations.

Moor House–Upper Teesdale, Cumbria and Durham, England (Lat 2° 20'W; Long 54° 40'N)
Sponsors: Natural Environment Research Council and English Nature

This site, which is also a UNESCO Biosphere Reserve, comprises two National Nature Reserves (NNRs), is situated in the northern Pennines, and is the largest and highest site presently in the network.

Moor House was designated as the first English NNR in 1952, and was intended both to protect rare or endangered communities of animals or plants and as a good example of an upland ecosystem to be used as an 'open-air' laboratory. The Reserve covers approximately 3500 ha and its altitude ranges from 290 m to 848 m. The geology is Carboniferous and consists of alternating strata of limestone, sandstone and shale into which dolerite of the Great Whin Sill has intruded. The eastern side of the Reserve slopes gently and rocks are commonly masked in clayey glacial till, making the drainage poor and resulting in the development of blanket bog with peat 2–3 m deep. The vegetation here is dominated by cottongrass and sphagnum moss (*Sphagnum* spp.), with heather at lower altitudes but not at higher altitudes. The western side of the Reserve slopes steeply and rocks are commonly exposed, giving rise to a wider variety of soils and more diverse vegetation. The climate has been described as sub-arctic oceanic with rainfall averaging 1900 mm per year and ground frost being recorded in all months of the year.

Upper Teesdale NNR was established in 1963, and extended in 1969, to protect its unique arctic/alpine plant communities and other flora and fauna. It covers 3000 ha and ranges in altitude from 300 m to 780 m. At the lowest altitudes there are deciduous woodlands and herb-rich meadows but the majority of the Reserve is used for sheep grazing and grouse shooting. The geology, soils and vegetation of much of the site are similar to those of Moor House NNR but an unusual feature is the 'sugar limestone' soil, derived from limestone which has been metamorphosed by the dolerite intrusion – this soil and the damp river-side soils support many of the rarer plant species.

Meteorological recording began at Moor House in the 1930s and at 560 m this was by far the highest point in Britain at which records had been made, excepting Ben Nevis where the station had closed at the turn of the century. Research began in earnest in 1952 with the opening of the Moor House Field Station. Studies on the impact and potential of moorland management led to the establishment of experimental plots on a variety of vegetation types to examine the impact of sheep grazing intensity on vegetation and soils. Plots were also established where heather burning for grouse management and tree planting took place. Moorland drainage ditches were studied on a small catchment basis. Annual variations in invertebrate populations, accumulation and erosion of peat, and inter-actions between terrestrial and aquatic ecosystems were also examined.

In the late 1960s and early 1970s the blanket bog ecosystem at Moor House was studied intensively as part of the International Biological Programme. This work obtained estimates of primary and secondary biological production and quantified the main pathways and rates of circulation of dry matter and nutrients. A major series of faunal studies, some of them long-term and carried out principally through the University of Durham, have contributed significantly to the selection of some variables to be measured in ECN. At this time intensive work was undertaken at Upper Teesdale, stemming from concerns over building the Cow Green Reservoir, particularly its impact on fish populations and terrestrial plant communities. A weather station was set up to determine whether the presence of the reservoir would have an impact on local meteorological variables. In the 1980s research emphasis moved to the deposition of atmospherically transported pollutants, including the effects of altitude and the role of clouds in deposition.

The site has been a focus in NERC's Terrestrial Initiative in Global Environmental Research (TIGER). Sampling for the programme has been distributed across the site and there are three major research locations. An altitudinal sequence of sites has been established in which global warming has been simulated by moving soil cores from high to low altitudes, with some being retained at the higher site to act as controls; vegetation and nutrient dynamics of the cores are being investigated. At a second location small, open-top, plastic greenhouses have been installed at the interface between bracken and heather to study the impact of increased temperature on competition between these species. Similar greenhouses are being used at a third site to study the impact of elevated temperature on invertebrate populations.

A meta-database is available, giving details of past and present research and an archive of research data and documents is also maintained.

North Wyke, Devon, England (Lat 3° 54'W; Long 50° 46'N)
Sponsor: Biotechnology & Biological Sciences Research Council

North Wyke Research Station, part of the Institute of Grassland and Environmental Research (IGER), lies in undulating countryside to the north of the Dartmoor National Park boundary, 6 km north-east of Okehampton. The River Taw flows northwards through the site. Altitude ranges from 120 m to180 m. Annual mean rainfall (1961–90) is 1034 mm, with a pronounced winter maximum from October to January. The wettest and driest months are January (129 mm) and July (53 mm) respectively. February (4.3°C) is the coldest month and July (15.3°C) the warmest. The annual total of hours of bright sunshine is 1448, with a mean daily range from 1.6 hours in December and January to 6.4 hours in June.

The site comprises 250 ha, of which 200 ha are lowland grassland typical of conditions in wetter, western Britain and 50 ha are deciduous woodland, mainly oak, ash and birch. The soils are predominantly impermeable clays and silty clays of the Hallsworth, Halstow and Denbigh/Cherubeer soil series, collectively known as the Culm Measures, and typically support permanent grass and dairying. The only well-drained soils on the Station comprise a narrow band of the Teign soil series alongside the River Taw.

IGER has a broad-based interest in climatic change with expertise spanning molecular biology, acclimatory physiology, environmental impact, and grassland agronomy and ecology. Specific topics include plant responses to drought, temperature extremes and elevated CO_2 levels. At North Wyke, which has new laboratory and extensive field-based facilities, there are research programmes on low-input grassland farming and relationships between grassland and the environment. Studies of nutrient cycling, especially of nitrogen but also of phosphorus, through the soil/plant/animal/farm wastes chain and effects on environmental quality, constitute major research programmes. There is particular emphasis on understanding the fluxes of nitrogen to waters as nitrate, and to the atmosphere as ammonia, nitrous oxide and nitrogen gas. Other research seeks to quantify and understand the emission of important greenhouse gases such as methane and nitrous oxide from grassland; pests and diseases, vegetation management to encourage species diversity, plant/animal interactions and the functioning of managed and natural grassland, and agroforestry are other research topics. About one-third of the managed land area at the site is sown to grass/clover swards reflecting the interest in lower-input managed grasslands.

North Wyke has records of temperature, rainfall and sunshine for the 30-year standard period 1961–90 with open-pan evaporation measurements available since 1983.

The TSS is a 0.66 ha paddock of old, biodiverse permanent grassland, comprised mostly of rough meadow-grass *(Poa trivialis)*, perennial rye-grass, Yorkshire-fog *(Holcus lanatus)* and creeping bent *(Agrostis stolonifera)* with a considerable ingress of rush *(Juncus)* species in one area, which receives no nitrogen fertilizer inputs. There has been no history of N fertilizers being applied to the TSS for many years, possibly since the 1940s. Controlled cattle grazing, with zero N input, is by beef cattle from April to October, their numbers being varied to maintain a compressed sward height of 5–6 cm. Sward productivity is measured under exclusion cages four times each year and liveweight gain of the grazing cattle is monitored annually.

Porton Down, Wiltshire, England (Lat 1° 43'W; Long 51° 07'N)
Sponsor: Ministry of Defence

Porton Down lies to the north-east of Salisbury, on the southern edge of Salisbury Plain and has been part of a Government establishment since the First World War when chemical weapons were first used on the battlefield. The requirement for a testing and experimental ground resulted in the setting up of the first restricted area at Porton. By 1918 the 'Range' had been acquired, encompassing approximately the same area now owned by the Chemical and Biological Defence Establishment (CBDE).

The continual use of the site for experimental purposes has effectively excluded the modern agricultural development seen over the vast majority of southern England. This has resulted in the core of the area remaining as a unique snapshot of Wessex downland as it might have been centuries ago.

The ECN site, set up in 1994, has been a Site of Special Scientific Interest (SSSI) since 1977. The SSSI covers 1227.4 ha, of which approximately 190 ha are beech and conifer plantation, 30 ha mixed broadleaved woodland, and the remainder semi-natural chalk grassland with successional scrub. Apart from Salisbury Plain, which lies only a few kilometres away, this is the largest remaining tract of chalk grassland in the British Isles. Typically, both of these military areas are surrounded by intensive, largely arable, agriculture.

Porton Down is underlain by Upper (Cretaceous) Chalk with a few small patches of clay-with-flints on the high ground mainly to the west. The soils are generally referable to the Icknield series which comprises shallow, mostly humose, well-drained calcareous soils. Deeper flinty calcareous silts can be found in small coombes and valleys. The only acid soils are the caps on high ground and a few small areas on valley-sides. These soils are almost all free-draining.

The climate lies between the strong maritime influence of the Atlantic and the 'continental' influence of south-east England (the sub-Atlantic). The site is within the region which experiences the highest maximum temperatures in the UK, but also some of the lowest minima. The coldest months are January and February (February mean temperature 3.2°C), and the hottest months are July and August (July mean temperature 15.7°C). The number of days with frost averages 109 whilst mean daily sunshine averages 1.7 hours in December and 7.0 hours in June. Rainfall is highest in autumn and winter due to the maritime influence. Mean rainfall is 750 mm (800–850 mm on the highest parts of the SSSI). Much of the site lies around the 125 m contour line although the highest points are at 170 m.

The vegetation has only relatively recently been well documented, with systematic surveys in the early 1970s and in 1991. The greatest interest lies in the tracts of high-quality semi-natural chalk grassland together with large populations of localised plants and exceptional lichen communities. The rarest forms are those of sheep's fescue/mouse-ear hawkweed/thyme *(Festuca ovina/Hieracium pilosella/Thymus* spp.), which have superb communities of terricolous lichens and abundant bryophytes. This grassland type is known only from Porton Down and from the Breckland of East Anglia. The main reason for the uniqueness of the downland vegetation is the continued existence of high populations of rabbits *(Oryctolagus cuniculus)*. Rabbits appear to have made a sustained recovery from myxomatosis at Porton Down and essentially 'drive' the dynamics of the grassland communities.

Apart from the woodland which is mostly recent and therefore largely lacking in botanical interest, the other feature of great importance is the juniper *(Juniperus communis)* scrub. The population of junipers, many of which are very old, is considered to be the finest in lowland Britain.

The animal populations are diverse and unique and reflect the rarity of the habitat Of the higher vertebrates, the stone curlew *(Burhinus oecdinemus)* population is of national importance (about 10% of the UK total). Among the invertebrates, the butterfly populations are of immense national importance and include such rarities as the silver-spotted skipper *(Hesperia comma)*. There are very large numbers of other species, particularly dark-green fritillary *(Mesoacidalia aglaja)* and marbled white *(Melanargia galathea)*. Other important invertebrate communities include those associated with juniper, as well as spiders and moths.

Rothamsted, Hertfordshire, England (Lat 0° 22'W; Long 51° 49'N)
Sponsor: Biotechnology & Biological Sciences Research Council

Rothamsted was founded in 1843 and is the oldest continuously functioning agricultural research station in the world. The estate lies to the west of Harpenden, 40 km NNW of London, on a gently undulating plateau 95–134 m in altitude. The soils are well-drained to moderately well-drained flinty silty loams over clay-with-flints and/or chalk. Rothamsted farm and most of the surrounding area have been taken up with arable agriculture for 2000 years.

A number of long-term experiments, the 'Classical Experiments', were established at Rothamsted in the second half of the 19th century. These cover cereal growth, grassland management and woodland regeneration, and are a unique record of environmental change; they are backed up by long runs of meteorological, soil chemical and biological, and plant composition data.

Broadbalk, the most famous of the Classical Experiments, was first sown to winter wheat in 1843 and harvested in 1844. Wheat has been grown on all or part of the field every year since then to compare different organic manures with inorganic fertilizers. The treatments were varied in the first few years but since 1852 a more permanent scheme has been established. Some changes have since been made to ensure that the experiment remains relevant to modern agricultural and environmental problems, including the use of modern varieties of wheat. Yields of grain and straw have been recorded yearly and samples taken for chemical analysis. These samples, together with those from the other Classical Experiments, are archived and are still available for analysis.

The Broadbalk soil, which is a heavy clay loam overlaying chalk (Batcombe series), has been sampled and archived irregularly since the start of the experiment, but in recent years a scheme to sample and store it on a five-year cycle has begun. The samples have been used to look at changes in soil organic matter pre- and post-atomic levels of radionuclides, the build-up of various heavy metals and increases in organic compounds such as dioxins and PCBs.

The Park Grass Classical Experiment was started in 1856, but it is thought that the field had previously been in pasture for up to 200 years. It was established to examine the effect of different fertilizers on the production of hay but soon developed into an experiment that shows the effect of different types and amounts of fertilizer and lime on botanical populations. The plots are cut for hay in June and again in October for silage.

Small amounts of lime were applied to Park Grass twice in the 1870s and 1890s; a regular liming scheme was introduced in 1903, with lime added to half of most plots every four years. This continued to 1964 when the plots were further divided into four subplots of which three are now limed to a pH of approximately 7, 6 and 5 respectively; the fourth is left unlimed. The overall pH of the experiment was 5.7–5.8 in 1856. Where fertilizer is added as ammonium sulphate the plots have become very acid, with the pH of the surface soil on some plots being as low as 3.9 (in water) where lime has not been added. The steady acidification has resulted in aluminium, suddenly released from the soil when the pH reaches a critical value of about 4.2, being taken up in large amounts by the hay. The fertilizer and lime treatments have also affected the species composition of the plots. The most acidic plots are dominated by one or two species, eg Yorkshire-fog and sweet vernal-grass (*Anthoxanthum odoratum*), and are similar to an upland grassland. The unfertilized plots are still the most diverse and similar to a lowland pasture, containing 40–60 species.

The whole site is encompassed by a loop of the River Thames, 5 km north-west of Oxford; it rises from an altitude of around 60 m on the river floodplain to 165 m at the top of Wytham Hill. A change in geology and soil type parallels the topography. Alluvium beside the Thames overlies Oxford Clay and away from the river the clay is exposed, with a large area of deep, heavy soils (eg Denchworth series) which are often waterlogged in winter. Towards the top of Wytham Hill is a thin band of sandstone, (giving rise to sandy Frilford series soil in places) and the summit is composed of coral rag limestone covered by extremely thin, well-drained soils (Sherborne or Morton series). A number of streams, many of which dry up in summer, rise on the estate and drain into the Thames.

The climate is typical of the Midland region; long-running data from the Radcliffe Meteorological Station in Oxford, where the first measurements were made in 1767, show a mean annual rainfall of 640 mm and mean air temperatures of 3.6°C for January and 16.4°C for July, the coldest and warmest months of the year respectively.

Vegetation on the estate was systematically surveyed in 1993 and 1994, building on earlier work. The ancient woodland, ie that believed never to have been cleared, is mostly abandoned coppice or coppice-with-standards. Hazel (*Corylus avellana*) and maple (*Acer campestre*) are the most common coppice species and oak the most frequent standard. Quite large areas of the woods are known to have grown up naturally in the last 200 years or so, after the abandonment of wood pasture, pasture, or cultivation, and the most prominent trees here are ash and sycamore. Neither the ancient nor the secondary woodland is actively managed and where possible dead wood is left where it falls. Plantations of various broadleaved and coniferous species have been established, most of which are 40–50 years old, though there are also some much older stands of beech. The semi-natural grasslands are mainly on the limestone areas and include areas of both ancient and modern origins; the latter have been the subject of a number of studies in vegetation succession. The agricultural land includes improved permanent pasture, grass leys and arable land used for wheat, barley, oilseed rape and other conventional crops; there are also hedgerows and some small wetland areas around drainage systems.

The animal populations include large numbers of fallow (*Dama dama*) and muntjac *(Muntiacus reevesi)* deer which have a major influence on woodland structure by browsing the shrub layer. Badgers (*Meles meles*) are also very common and have been studied in great detail; wood mice (*Apodemus sylvaticus*) and bank voles (*Clethrionomys glareolus*) have been monitored since 1948. As many as 152 bird species have been recorded on the estate, approximately half of which are regular residents. In-depth studies on blue tits (*Parus caeruleus*) and great tits (*Parus major*) have allowed numbers to be followed from year to year and have led to an understanding of some of the mechanisms determining breeding success in relation to climatic fluctuations. Birds of agricultural land, such as starlings (*Sturnus vulgaris*), have also been the subject of much intensive

26

research in recent years. Amongst the invertebrates the site is notable for the presence of five species of hairstreak butterflies (*Strymonidia* spp.) and important research on many other groups has been carried out here.

Y Wyddfa/Snowdon, Gwynedd, Wales (Lat 4° 05'W; Long 53° 04'N)
Sponsors: Countryside Council for Wales & Welsh Office

The Y Wyddfa/Snowdon site is situated 19 km south-east of Bangor in north Wales. The site is located in very rugged terrain and includes the summit of Y Wyddfa or Snowdon, the highest mountain in England and Wales. Its altitudinal range is 298–1085 m and in addition to Y Wyddfa it contains three other summits over 800 m. Snowdon is a popular destination for walkers and the visitor pressure, though high, is generally concentrated along the main footpaths. The site is part of the Y Wyddfa/ Snowdon National Nature Reserve and is managed by the Countryside Council for Wales under agreement with the owner.

The bedrock is a mixture of Ordovician acidic and basic volcanic rocks, with localised igneous intrusions. Evidence of glaciation is widespread and there is a prominent set of corrie moraines dating from the Loch Lomond Stadial. There are five lakes within the site, the three largest of which form a 'staircase'. The soils are varied and include brown earths, brown podzolic soils, gley podzols, gleys, stagnohumic gleys, organic peat soils and humic rankers. There are significant areas of frost-shattered scree beneath the steeper slopes.

Rainfall data have been collected at the site since the early part of the 20th century and other meteorological variables have been measured systematically since the 1960s. Rainfall is high and mean annual range across the site is generally in the range 3000–4000 mm. The mean annual number of days with rainfall >1 mm is 200 and, on average, snow lies for between 16 and 100 days per year, depending on altitude and aspect. Mean monthly air temperature at the TSS ranges from 2.8°C (February) to 15.3°C (August).

The predominant vegetation is acidic grassland with large areas dominated by sheep's fescue and bent in more freely drained areas and mat-grass where drainage is somewhat impeded. There are localised areas of calcicolous grassland with wild thyme (*Thymus praecox*), red fescue (*Festuca rubra*) and marsh thistle (*Cirsium palustris*). The largest of these areas includes the ECN TSS. There are small areas dominated by heather on steeper ground. Where the cliffs have a base-rich influence, tall-herb and crevice vegetation has developed; these locations are an important locus for a number of arctic/alpine species including the Snowdon lily (*Lloydia serotina*) which in Britain is restricted to the mountains of Snowdonia. Scree vegetation occurs beneath the major cliffs, with frequent parsley fern (*Cryptogramma crispa*). The site also contains the largest stand of vegetation dominated by dwarf juniper (*Juniperus communis alpina*) south of the Scottish Highlands. Finally, in the lower part of the site there are areas of blanket mire with abundant cottongrasses.

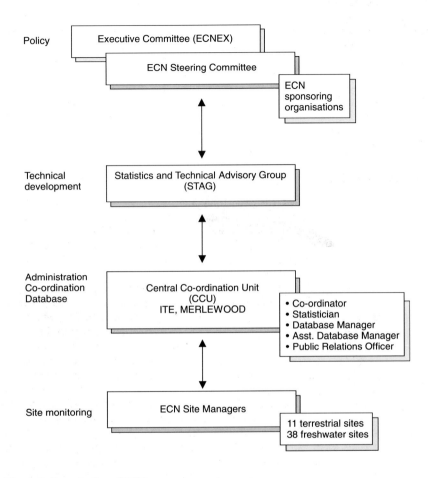

Policy

Executive Committee (ECNEX)

ECN Steering Committee

ECN sponsoring organisations

Technical development

Statistics and Technical Advisory Group (STAG)

Administration Co-ordination Database

Central Co-ordination Unit (CCU) ITE, MERLEWOOD

• Co-ordinator
• Statistician
• Database Manager
• Asst. Database Manager
• Public Relations Officer

Site monitoring

ECN Site Managers

11 terrestrial sites
38 freshwater sites

Figure 2. Organisation of ECN

The site is unenclosed and is grazed by Welsh mountain sheep, in addition to which there is a small herd of feral goats.

Research at the site dates back to the 1950s when work started on the distribution patterns of sheep grazing on different vegetation types. A number of small sheep exclosures which were set up in the 1950s remain on the site and form part of a sequence of such sites extending across the mountains of north Wales. Research into the energy flow and productivity of calcicolous grassland was undertaken as part of the International Biological Programme during the 1970s. Research has been undertaken more recently to assess the effects of different grazing levels on invertebrate populations.

Operation of ECN

ECN operates by consensus of its participating agencies, each of which is represented on the ECN Steering Committee, the body responsible for the main policy decisions affecting the network. The Steering Committee normally meets annually but a subgroup, the ECN Executive Committee (ECNEX), expedites urgent decisions as well as developing ideas and recommendations for the Steering Committee. The two Working Groups which were instrumental in developing the technical and statistical elements of the network have recently been amalgamated to form a joint Statistics and Technical Advisory Group (STAG) which reports to the Steering Committee. NERC provides the day-to-day management of the network by providing and supporting the ECN Central Co-ordination Unit (CCU) which is responsible for standardising procedures and for co-ordinating data collection and management. The CCU has four full-time staff: the ECN Co-ordinator, a statistician, a data manager and an assistant data manager, plus a part-time public relations officer, all of whom are staff members of the Institute of Terrestrial Ecology. At each site a sponsoring agency provides a Site Manager responsible for organising the timely collection and initial processing of data according to the agreed Protocols, and transmission of the data to the ECN Data Manager in an agreed format. The current organisation of ECN is shown in Figure 2.

J.M. Sykes
ECN Co-ordinator 1992–95

References Bunce, R.G.H., Barr, C.J., Clarke, R.T., Howard, D.C. & Lane, A.M.L. 1996. Land classification for strategic ecological survey. *Journal of Environmental Management*, **47**, 37–60.

Cormack, R.M. 1994. Statistical thoughts on the UK Environmental Change Network. In: *Statistics in ecology and environmental monitoring*, edited by D.J. Fletcher & B.F.J. Manly, 159–172. Dunedin, New Zealand: University of Otago Press.

Fritz, J.-S. 1991. *A survey of environmental monitoring and information management programmes of international organisations.* Munich: UNEP-HEM.

Heal, O.W. 1991. The role of study sites in long-term ecological research: a UK experience. In: *Long-term ecological research*, edited by P.G. Risser, 23–44. Chichester: Wiley.

Heal, O.W., Menaut, J.C. & Steffen, W.L., eds. 1993. *Towards a Global Terrestrial Observing System (GTOS): detecting and monitoring change in terrestrial ecosystems.* (MAB Digest 14 and IGBP Global Change Report no. 26.) Paris: UNESCO and Stockholm: IGBP.

Hinds, W.E. 1984. Towards monitoring of long-term trends in terrestrial ecosystems. *Environmental Conservation,* **11**, 11–18.

House of Lords Select Committee for Science & Technology. 1984. *Agricultural and environmental research.* (4th Report.) London: HMSO.

International Geosphere-Biosphere Programme. 1992. *Global change and terrestrial ecosystems – the operational plan.* (Report no. 21.) Stockholm: **IGBP.**

Likens, G.E. 1983. A priority for ecological research. *Bulletin of the Ecological Society of America,* **64**, 234–243.

McCormack, P.S. 1990. *Environmental monitoring systems in the United Kingdom: towards an integrated approach.* Internal DOE Working Paper presented to UNECE Workshop, Geneva.

Natural Environment Research Council. 1976. *Biological surveillance.* (NERC Publication Series B, no. 18.) Swindon: NERC.

Natural Environment Research Council. 1986. *Report of the Working Group on Long Term Reference Sites.* (Unpublished report.) Swindon: NERC.

Strayer, D., Glitzenstein, J.S., Jones, C.G., Kolasa, J., Likens, G.E., McDonnell, M.J., Parker, G.S. & Pickett, S.T.A. 1986. *Long-term ecological studies.* New York: Institute of Ecosystem Studies.

Tinker, P.B. 1994. Monitoring environmental change through networks. In: *Long-term experiments in agriculture and ecological sciences,* edited by R.A. Leigh & A.E. Johnston, 407–420. Wallingford: CAB International.

Woiwod, I.P. 1991. The ecological importance of long-term synoptic monitoring. In: *The ecology of temperate cereal fields,* edited by L.G. Firbank, N.Carter, J.F. Darbyshire & G.R. Potts, 275–304. Oxford: Blackwell Scientific.

Chapter 2   ECN CORE MEASUREMENT PROTOCOLS

The Protocols for the ECN core measurements contain detailed information on the procedures to be followed for sampling, treatment of samples and data collection. Testing of the practicality and utility of the procedures has taken place over a three-year period (1993–96), as a result of which some modifications have been. The methods can therefore be regarded as tried and tested but there is a continuing programme of appraisal which will almost inevitably lead to future minor changes.

The arrangement of the Protocols follows a logical order (see Table 2), starting with those which relate to initial set-up and information on continuing site management; driving variables are followed by state and response variables; finally there are two Protocols which relate to the handling and chemical analysis of waters collected during the operation of several other preceding Protocols.

Almost all the core measurements have been designed to be used at all ECN's terrestrial sites, but with the following exceptions.

- The water discharge (WD Protocol) and water quality (WC Protocol) measurements are undertaken only where the whole catchment area of a natural, perennial stream lies within the site.

- The moorland breeding birds (BM Protocol) are recorded only at upland moorland sites, and at such sites the Protocol for common breeding birds (BC) is not followed.

- At sites which are intensively grazed by sheep, the assessment of rabbit populations (BU Protocol) using dropping counts is impracticable.

- At sites where there are no ponds or ditches, it will be impossible to record the phenology of frog spawning (BF Protocol).

Some parts of a Protocol may be inappropriate to a particular site, as is the case with the vegetation Protocol (V). Thus, only the VB, VC and VF Protocols are mandatory at all sites, and, whilst it is likely that the Protocol dealing with linear features (VH) will also be appropriate at most sites, those dealing with woodlands (VW), permanent pastures (VP) and cereals (VA) will be used at fewer sites.

There are apparent inconsistencies between some Protocols which result from their having been derived from other, already established methods. Thus, the random kilometre squares used in the breeding birds Protocol (BB) and the subjectively selected kilometre squares used in the bat Protocol (BA) follow existing methods in the sectoral programmes of other organisations. ECN has standardised where possible on Greenwich Mean Time (GMT), although some already established methods use British Summer Time (BST).

Table 2.   Summary of core measurements for ECN terrestrial sites

Measurement	Frequency	Protocol	Page
Location & marking of TSS	Start of programme	LM	33
Land use & site management	As necessary	LU	35
Meteorology			37
Automatic weather station	Hourly averages or totals	MA	38
Standard observations	Daily (or weekly by default)	MM	45
Atmospheric chemistry: NO$_2$	Tubes changed 2-weekly	AC	47
Precipitation chemistry: major ions	Weekly analysis	PC	53
Hydrology			
Surface water discharge	Continuous	WD	60
Surface water chemistry & quality: major ions	Weekly analysis	WC	64
Soils			
Survey at 1:10 000 or 1:25 000	Start of programme	SB	66
Core sampling: major elements	5-yearly	SF	69
Pit sampling: heavy metals & physical properties	20-yearly	SC	70
Soil archiving	5-yearly	SA	73
Soil solution chemistry			
Major ions	2-weekly, Oct–May	SS	75
Vegetation			
Baseline survey	Start of programme	VB	88
Coarse-grain sampling	9-yearly	VC	90
Fine-grain sampling	3-yearly	VF	90
Woodlands	9-yearly	VW	91
Hedgerows & linear features	3-yearly	VH	93
Permanent pasture	4 x per year	VP	93
Cereals	Annually	VA	94
Invertebrates			
Tipulids	April & Sept	IT	102
Moths	Daily	IM	104
Butterflies	Weekly, April–Sept	IB	109
Spittle bugs	Annually	IS	111
Ground predators	2-weekly, May–Oct	IG	118
Vertebrates			
Birds			
Common breeding birds	10 x per year, March–June	BC	125
Moorland breeding birds	2 x per year, April–June	BM	128
Breeding bird survey	2 x year, April–June	BB	131
Bats	4 x per year, June–Sept	BA	134
Rabbits & deer	2 x per year, March & Sept	BU	139
Frog spawn	Weekly in spring	BF	141
Initial water handling		WH	145
Analytical guidelines for waters		AG	149

LOCATING AND MARKING THE SITES

Aim *To locate the position of and to mark the Target Sampling Site (TSS) and the Sampling Site (SS)*

Rationale At each ECN site an area of 1 ha, preferably a square 100 m x 100 m, is required for sampling, which for practical reasons cannot be more widely dispersed. This is the Target Sampling Site (TSS) and destructive sampling within it should be kept to a minimum. The TSS is the central hectare in a square block designated as the Sampling Site (SS). Where space permits, the SS should cover 9 ha and have soil, vegetation and management similar to the TSS. The SS is used for destructive sampling and for research related to ECN monitoring. Dispersed monitoring, eg of vegetation, should include the TSS.

Method **Location**

The TSS is intended to be representative of the major or predominant vegetation type, soil and management of the ECN site and it should be subjected to consistent management over the life of the ECN programme. It should, where possible, have uniform soil and vegetation. The proximity of outside influences such as high hedgerows, which may affect the uniformity of the TSS, should be avoided. It is recognised that, at some ECN sites where management is intensive, there are severe restrictions on the choice of an area for use as a TSS.

Procedure

It is recommended that the axes of the TSS should be aligned with the axes of the Ordnance Survey National Grid, and should preferably correspond with the 100 m grid. In any case the National Grid references of the TSS corners should be recorded with the greatest possible precision. Thereafter a local six-figure numeric co-ordinate system should be used to reference points or areas within the SS and TSS, using the system shown in Figure 3. The origin of the local system is in the south-west corner of the SS. This system allows points or quadrats to be referenced with a resolution of 1 m; where quadrats are to be referenced, the local co-ordinates of the south-west corner of the quadrat should be used and the dimensions of the quadrat stated.

Permanent marking of the SS and TSS is essential and can take various forms, such as concrete or stone blocks sunk into the ground and either flush with the ground surface or raised above it. Subject to avoiding the use of materials likely to affect soil and vegetation, the material chosen should be that most appropriate to the particular site. It may be helpful to mark permanently each

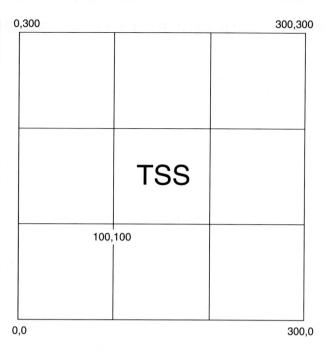

0,300   300,300

TSS

100,100

0,0   300,0

Figure 3. System for referencing points or areas within the SS

corner of the TSS, and preferably the SS, with the six-figure reference of the local numeric co-ordinate system.

Within the SS or TSS it may be convenient to have other markers arranged in a systematic fashion which help to locate points or areas at which sampling has occurred. These should be marked with a plate carrying the local numeric co-ordinates.

Detailed records must be kept of the location of all scientific work undertaken on the TSS and SS or any form of disturbance, such as wind damage or erosion.

Author   *J.M. Sykes*

LAND USE AND SITE MANAGEMENT

Aim   *To record location, timing and intensity of site management activities*

Rationale   In long-term monitoring studies, changes in land management can confound trends in measurements resulting from changes in extraneous factors such as climate and pollutants. Management should therefore be consistent from year to year, at least at the ECN Sampling Site (see LM Protocol). Ideally, management will be 'typical' of the management of similar ecosystems for the geographical area in which the ECN site lies. The ECN Site Manager should make every effort to ensure that the Sampling Site is managed in a consistent way.

Method   Detailed records of management activities at the ECN site, and particularly on the Sampling Site, should be maintained for incorporation within the ECN database. Where sites already have a detailed recording system for management, information about the system should be sent to the ECN Central Co-ordination Unit.

ECN sites have been selected to cover a wide range of ecosystems and therefore a wide variety of management activities are practised within them. It is therefore difficult to provide formal categories appropriate to all sites. However, Site Managers will be aware of the principal management activities occurring at their particular site. It is essential to record location, timing, intensity and method of the applied treatment or management. The most appropriate basic management units, eg fields or compartments, should be identified and coded on to a baseline map for consistent recording over the duration of the ECN programme. Records of management activities should include their map unit code, and the date or date range to which they relate. The map will form the basis for handling these management records within the ECN geographical information system (GIS).

Sampling Site

As much detail on management activities as possible should be recorded for the Sampling Site. Examples are woodland thinning, spreading fertilizer, introducing farm stock, harvesting and heather burning. When appropriate, the rates of these activities should be included, eg the dose rate for fertilizer, the basal area of trees removed, the number of cattle brought in to graze. Traumatic natural phenomena, eg flooding or woodland wind-throw, should also be recorded. The extent of a management activity or natural phenomenon should be recorded on a map of the Sampling Site. To simplify digitising these maps, a standard base map of the Sampling Site should be used on each recording occasion. This map should be large scale (1:2500 or greater) and should include the locations of the subplots used for the various ECN measurements.

ECN site

Some ECN recording, eg of vegetation, takes place over the whole ECN site. Annual records of land use and summary information about stocking rates and chemical applications should be maintained at all ECN sites, but where intensive agriculture is practised recording should be more frequent so that the full annual cycle of land use is recorded. Air photographs should be used as historic records of land use and also to plot precisely important features not shown on published maps, eg new field boundaries. Additional maps should be made to show traumatic natural phenomena and air photography may also be of helpful. Again, to simplify digitising these maps, a standard base map of the site should be used on each recording occasion. On sites smaller than 25 km^2, the resolution should be at a scale of 1:10 000 and on sites greater than this it should be at 1:25 000.

Author   *J.K. Adamson*

METEOROLOGY

Overall aim   *Unattended and attended recording of meteorological data*

Rationale   Climate is not constant: records over the last two centuries have
shown that both long-term trends and short-term perturbations in
climate can occur. There is much present concern that human
activities may be inadvertently changing the earth's climate
through an enhanced 'greenhouse effect', by continuing emissions
of carbon dioxide and other gases which will cause the
temperature of the earth's surface to increase. The local impact of
these global changes in climate is not yet known. Climatic
conditions, particularly temperature and rainfall, probably provide
the most important constraint on ecological processes; a
knowledge of long-term changes in climate therefore provides the
starting point for any analysis of changes in ecosystem structure
and dynamics at a site.

AUTOMATIC WEATHER STATION

Aim   *Unattended recording of meteorological data*

Rationale   Several ECN sites are geographically remote and this severely restricts the possible frequency of meteorological measurements because these sites cannot be visited consistently at intervals of less than a week. Meteorological data collected at frequent intervals over very long periods are needed both for many ECN purposes and also for environmental research programmes associated with ECN. During the last 20 years there have been considerable advances in the technical development of sensors and data logging devices for automatically measuring and recording meteorological data. Equally important has been the increased reliability of these sensing and logging devices. ECN automatic weather stations (AWS) have therefore been configured to measure and record a basic set of meteorological variables which will:

- provide information on temporal changes in these variables,
- aid in the interpretation of other ECN core measurements,
- provide data to support other environmental research at ECN sites.

It is strongly recommended that wherever possible duplicate stations should be installed to measure significant environmental gradients where these may exist within an ECN site.

Method   **Equipment**

The need to standardise equipment across the network requires the installation of an instrument configured to ECN specifications. The following variables will be recorded at each site at 5-second intervals and stored as hourly summaries on the hour, with the accuracy and resolution given below in brackets:

Wind speed ($\pm1\%$; 0.1 m s^{-1})
Wind direction ($\pm2°$; 1°)
Wet bulb temperature in non-aspirated screen ($\pm0.1°$C; 0.1°C)
Dry bulb temperature in non-aspirated screen ($\pm0.1°$C; 0.1°C)
Kipp solarimeter ($\pm1\%$; 1 W m^{-2})
Net all wave radiation ($\pm5\%$; 1 W m^{-2})
Tipping bucket raingauge (unheated), placed on paving slab, in a ground level pit at exposed upland sites ($\pm5\%$; 0.5 mm)
Soil temperature, 10 cm under bare soil ($\pm0.1°$C; 0.1°C)
Soil temperature, 30 cm under grass ($\pm0.1°$C; 0.1°C)
Downward-facing silicon cell, for albedo/snowcover($\pm5\%$; 1 W m^{-2})
Gypsum block, 20 cm under grass, for soil moisture measurement ($\pm10\%$; 0.01 bar)

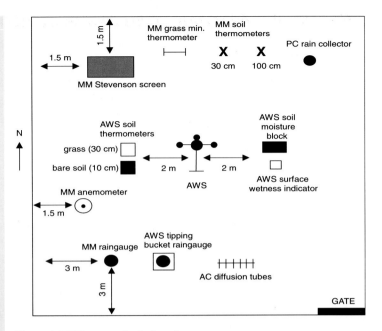

Figure 4. ECN meteorological enclosure

Surface wetness indicator (indicating the number of minutes per hour during which the surface is 'wet') at ground level (±1 min; 1 min)

An independent copy of the logger programme should be kept on site. Raingauges placed in pits should be surrounded by an anti-splash grid.

Location

The AWS is to be sited according to British Meteorological Office site requirements (Meteorological Office 1982), within 500 m of the TSS and close to the manual meteorological instruments (see Figure 4). If for operational reasons the AWS has to be sited more than 500 m from the TSS, this should be agreed in advance by the ECN Central Co-ordination Unit. The maximum permissible distance of the AWS from the TSS is 2 km.

Operation

Installation of the AWS should follow the instructions in Appendix I and Plate 1. Data are to be downloaded to a PC fortnightly, *via* a storage module, and the station checked according to the procedures provided in Appendix II. The station should be serviced and the sensors calibrated annually. Further notes on maintenance of equipment are also provided in Appendix II. Data quality control should be carried out according to Appendix III.

Plate 1. The automatic weather station

Time 4 hours/month (field + computer storage)

Authors *T.P. Burt and R.C. Johnson*

MA Protocol

Appendix I. Procedure for installing an automatic weather station (AWS)

Procedure

1. Dig a hole large enough to accept the main pole base and approximately 40 cm deep. A concrete base is not normally required unless the soil is particularly unconsolidated.

2. Fix the lower half of the main pole on to its base and lower it into the hole. Keep the pole vertical, using a spirit level, whilst infilling the hole.

3. Lower the bottom cross-arm on to the lower half of the main pole.

4. Assemble the solarimeter to the upper half of the main pole using the captive bolt which is in the solarimeter housing. Connect the cable to the solarimeter.

5. Slide the top cross-arm on to the upper half of the main pole and tighten the U-bolt to hold it in place.

6. Fix the top half of the main pole to the lower half.

7. Adjust the height of the top cross-arm to be 2 m above the ground level and oriented east/west.

8. Fix the guy-wires to the top cross-arm and to stakes in the ground, then adjust them so that the main pole is vertical.

9. Slide the lower cross-arm up the main pole to 1 m above ground level and orientated north/south, and tighten the U-bolt to hold it in place.

10. Fit the wind vane, anemometer, temperature screen and sensors, and net radiation sensors to the poles as shown in Plate 1 and connect the cables.

11. Place the raingauge on its concrete base, level the base, using the spirit level in the gauge, by adjusting the feet and connect the cable.

12. Fit the main loom junction box to the main pole using the jubilee clips provided.

13. Connect the sensor cables to the junction box in the correct order. Tape all cables tightly to the cross-arms and main pole.

MA Protocol

14. Fit the wick to the wet bulb sensor and fill the reservoir with distilled water.

15. Secure the solar panel to the main pole facing south.

16. Level all sensors and align the wind direction sensor to face north.

17. Install the logger in its housing and connect the main cable to the logger.

18. Hammer the earth rod into the ground and connect the earth-braided wire to the earth rod and stud on the logger.

19. Connect the solar panel cable to the logger.

20. Connect the storage module to the logger to download.

21. Close the logger.

Notes

The soil temperature probes should be installed to the west of the main mast, at least 2 m away from it or as far from it as the cable supplied will allow.

The albedometer should be mounted 1 m above the ground in a roughly SSE direction from the main mast, ie as close to south as possible without interfering with the net radiometer.

The soil moisture block should be installed to the east of the main mast, at least 2 m away from it or as far as the cable supplied will allow.

The surface wetness indicator should be installed on a representative piece of ground to the east of the main mast, at least 2 m away from it or as far as the cable supplied will allow.

The only parts not supplied with the AWS by the manufacturer are a concrete slab for the raingauge base and a suitable weatherproof housing for the logger.

Appendix II. Notes on maintenance of AWS equipment

1. The grid on the surface wetness indicator should be cleaned regularly, preferably on each fortnightly visit to the AWS, with a soft cloth and mild household detergent.

2. The wick on the temperature probe should be checked for dampness and contamination and be replaced if necessary; 3–12 month intervals are recommended depending on the environment. Wet bulb reservoirs should be topped up fortnightly with distilled water.

3. Net radiometer domes should be checked for cleanliness or damage every six months. The silica gel should be changed if necessary.

4. The solar dome should be cleaned fortnightly and the silica gel changed if necessary.

5. Gypsum blocks are destroyed by frost; under most circumstances likely to be met in ECN they should be unaffected at a depth of 20 cm, but should be replaced if there is a likelihood of there having been frost at this depth. They should in any case be replaced annually in spring.

6. The logger output should include a daily reading of battery voltage. Check that the battery voltage is not outside the range 11.5–14 volts. If the voltage falls outside the level recommended by the manufacturer, the batteries should be replaced. Batteries should in any case be replaced every two years. Note that an external battery must be connected to preserve the software when replacing the internal battery.

7. Check that the anemometer and wind vane rotate freely. Change the anemometer reed switches annually at exposed sites.

8. Check that the raingauge funnel and filter are free of debris, snow and ice. Clear any ice from the tipping buckets (without making a tip). Check that the base is level using a spirit level on the base, and adjust the feet if necessary.

9. Check that no cables are perished or damaged by animals and replace if necessary.

10. Check the effectiveness of the guy wires supporting the mast.

11. Check the security of the earthing cable and rod.

Appendix III. Procedure for quality control of AWS data

1. Check that the first and last data points agree with the times of data removal from the logger.

2. Check that the dry and wet bulb observations from the manual station agree with logger data, remembering that the logger gives a mean value for the last complete hour of recording.

3. Check that the rainfall total for the manual station agrees with amounts recorded by the logger by adding the daily logger totals to the hourly values for the days on which the logger has been downloaded.

4. If defective equipment has been noted during a field visit, identify erroneous data on the logger output and note the date ranges of errors for the information of the ECN Data Manager.

5. Check the output for missing data or unusual sequences of values, eg a series of constant values may appear when a variable series is expected. This may denote faulty sensors. If a sensor has been removed for repair, a missing data value, to be agreed with the ECN Data Manager, should appear in the logger output. Note any occasions when freezing may have affected sensors.

6. Make notes on the results of all checks, and any changes which are necessary or have been made and ensure that these notes accompany the data when they are transmitted to the ECN Data Manager.

STANDARD METEOROLOGICAL OBSERVATIONS

Aim

To conduct standard meteorological observations according to protocols laid down by the British Meteorological Office (BMO)

Rationale

In common with a large number of sites in the UK, all ECN sites have some historic meteorological data, many of which have been collected to meet the criteria of the Meteorological Office. In order to preserve the continuity of these historic data and to provide calibration and back-up for the ECN automatic weather stations, each site will have an array of standard meteorological instruments which will be read at weekly intervals, except where they constitute an existing Meteorological Office Climatological Station at which daily recording is the norm.

Method

Equipment

Each meteorological station will have the following items:

Large Stevenson screen with iron stand
Dry and wet bulb thermometers(±0.1°C; 0.1°C)
Maximum and minimum thermometers (±0.1°C; 0.1°C)
Grass minimum thermometer(±0.1°C; 0.1°C)
Soil thermometers at 30 cm and 100 cm (±0.1°C; 0.1°C)
Octapent raingauge, BMO pattern Mk2A
Run-of-wind counter anemometer, to BMO specification (unless an alternative anemometer is already available in addition to the AWS).

Where appropriate, certification of thermometers to BSI standard is provided by the supplier.

Location

The manual station should be sited alongside the AWS if possible (see Figure 4), though where there is an existing Meteorological Office Climatological Station it is accepted that this may not be adjacent to the TSS and its AWS.

Operation

Except where otherwise stated, methods follow the standards given in the *Observers' handbook* (Meteorological Office 1982) produced by the BMO. All instrumentation should be to the BMO standard. For Climatological Stations, BMO personnel undertake field calibration of instruments periodically.

Instruments at stations established for ECN purposes will be read weekly on Wednesdays at 0900 GMT. Daily (0900 GMT) readings will continue for existing BMO Climatological Stations and for these Stations a copy of the monthly data sheet, as submitted to the BMO (Metform 3208B), will be sent to the ECN Data Manager in addition to the agreed machine-readable data. Any queries raised subsequently by the BMO as a result of their quality control process should be reported to the ECN Data Manager. If sites are already recording sunshine hours, snow depth and soil temperatures for non-ECN purposes, these data should also be sent to the ECN Data Manager.

Time

1h/month for the weekly sites.

Supplier

Casella London Ltd
Regent House
Wolseley Road
Kempston
Bedford MK42 7JY, UK

Authors

T P Burt and R C Johnson

Reference

Meteorological Office. 1982. *Observers handbook.* 4th ed. London: HMSO.

ATMOSPHERIC CHEMISTRY

Aim *To detect and define changes in periodic mean atmospheric concentrations of nitrogen dioxide*

Rationale Natural and man-made sources emit oxides of nitrogen in approximately equal quantities, but, whilst the former tend to have equal world-wide distribution and are relatively constant over time, the latter are concentrated in or around centres of population. Nitric oxide (NO) and nitrogen dioxide (NO_2) are the most important oxides of nitrogen in urban atmospheres and it is NO_2 which has the more significant health and ecosystem effects. Major sources of NO_2 in urban areas result from fuel combustion in motor vehicles, power generation, heating plants and industrial processes, mostly by oxidation of NO emissions from these sources. NO_2 is a respiratory tract irritant, is toxic at high concentrations, and is involved in the formation of photochemical smog and acid rain. It can also cause direct damage to crops and other vegetation, together with SO_2 and ozone (UNEP/WHO 1994). High concentrations of NO_2 can have an indirect effect on ecosystems by providing an increased nitrogen input and this is important in the context of systems with a low natural nitrogen demand. The method selected for measuring nitrogen dioxide uses passive diffusion tubes; this method is well researched and reliable, though variation between individual tubes requires that they are replicated; the method has a low capital cost, requires no on-site power source and is reasonably inexpensive to operate.

The possibility of extending the ECN measurement programme to include other pollutant atmospheric gases, particularly ammonia, sulphur dioxide and ozone, has been considered and is under continuing review. At present, however, there appear to be no acceptably reliable passive methods.

Method **Equipment**

The construction of the diffusion tubes is shown in Figure 1. Discs of stainless steel mesh coated with NO_2 absorbent are secured in the upper end of the plastic tube. The 1 cm diameter discs of stainless steel mesh can be purchased ready-cut from the supplier of tubes and caps. Alternatively, ready-made tubes, which include prepared discs, may be purchased from the same source.

Where ready-made tubes and disks are not being used, equipment is prepared as follows. Tubes and caps are placed in a 5% acid wash overnight and then washed in Decon 90 for one hour and rinsed with distilled water. Stainless steel mesh disks should not be placed in acid but are stored overnight in Decon 90 and placed in an ultrasonic bath for 20 minutes; the disks are then rinsed

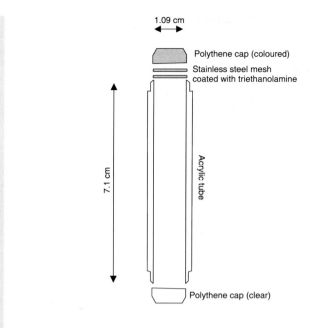

1.09 cm

Polythene cap (coloured)

Stainless steel mesh
coated with triethanolamine

7.1 cm

Acrylic tube

Polythene cap (clear)

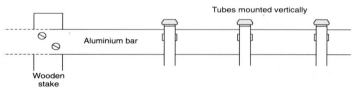

Tubes mounted vertically

Aluminium bar

Wooden
stake

Figure 5. Diffusion tube used for long-term monitoring of nitrogen dioxide, and its position in the field

thoroughly with distilled water and placed on filter paper to dry. Forceps should be used when handling the disks. Two dry disks are placed in each cap and 30 μl of a 10% aqueous solution of triethanolamine is pipetted into each cap. A tube is fitted into the coloured cap and the other end of the tube is then sealed with a white cap. Samples are stored in a refrigerator until they are needed. The diluted absorbent should be made up freshly shortly before use, and clean disks are best stored dry.

Location

Three diffusion tubes are mounted in clips (see Figure 5) which are either fastened to the pole of the bulk precipitation collector (see PC Protocol) or to a post at a height of 1.5 m above ground level. If a meteorological instrument enclosure exists, the post should be sited within the enclosure.

Sampling

The white cap is removed from each tube immediately prior to its deployment and the tubes are placed vertically with their open ends pointing downwards. The tubes are collected after two weeks and the white caps are re-fitted to the tubes before their removal from the post. Three blank tubes are transported to the site but are not exposed on arrival. They should be returned to the laboratory the same day, stored in a refrigerator during the two-week sampling period and analysed with the experimental tubes.

Labelling

Tubes should be labelled as follows:
- the ECN Measurement Code (AC),
- the ECN Site ID Code (eg 04 for Moor House),
- the Location Code (eg 01),
- the individual tube code (E1, E2, or E3 for exposed tubes; B1, B2, or B3 for blank tubes),
- the collection date ('Sampling Date') (eg 01-Jan-1996).

This unique reference MUST accompany the analytical results for transfer to the ECN database.

Authors *D. Bojanic, J.K. Adamson, A.P. Rowland and J.M. Sykes*

References **Hargreaves, K.H.** 1989. *The development and application of diffusion tubes for air pollution measurement.* PhD thesis. University of Nottingham.

United Nations Environment Programme/World Health Organization. 1994. *GEMS/AIR methodology reviews vol 1: quality assurance in urban air quality monitoring.* Nairobi: UNEP.

Appendix I. Analytical procedure

The absorbent is analysed to give the total NO_2 absorption for the sampling period and this is used to calculate mean NO_2 concentration over the same period

The operator must wear gloves to prevent contact with chemicals or solutions. The diffusion tubes must be rinsed immediately following analysis to reduce the risk of contaminating other handlers.

Reagents

Chemicals – AR Grade
Water – deionised or distilled

Sulphanilamide
HARMFUL IF SWALLOWED. AVOID CONTACT WITH SKIN AND EYES
Dissolve 10 g sulphanilamide in water, add 25 ml concentrated H_3PO_4 and dilute to 500 ml. Store in a refrigerator.

N-1-Naphthylethylene-diamine dihydrochloride (NEDA)
HARMFUL
Dissolve 0.14 g NEDA in water and dilute to 100 ml. Store in a refrigerator.

Reagent Y for samples
Mix 100 ml sulphanilamide, 100 ml water and 10.0 ml NEDA reagent. Prepare on the day of analysis.

Reagent X for standards
Mix 10.0 ml sulphanilamide with 1.0 ml NEDA reagent. Prepare on the day of analysis.

Stock standard A (2500 mg l^{-1} NO_2)
Dissolve 0.9375 g dried sodium nitrite in water and dilute to 250 ml in a volumetric flask.

Stock standard B (25 mg l^{-1} NO_2)
Dilute 1.0 ml of stock standard A to 100 ml in a volumetric flask.

Working standards (0.25, 0.75, 1.25 µg ml^{-1} NO_2)
Prepare fresh daily. Pipette 1.0, 3.0 and 5.0 ml aliquots of standard B to 100 ml in separate volumetric flasks and dilute to volume.

1. Standards: pipette 1.0 ml standard or blank into a tube. Add 1.1 ml reagent X.

Samples: remove the white cap from the tube and add 2.1 ml reagent Y.
Replace caps and shake.

2. Shake sample/standard after 15 and 30 minutes, and then record the absorbance at 542 nm against a water blank.

3. Calculate the weight of nitrite from the calibration curve and report the results, in µg, to three significant figures.

4. Nitrogen dioxide concentration in ppb can be calculated from the diffusion tube exposure time, tube dimensions and the amount of nitrite collected (Hargreaves 1989).

Appendix II   Equipment details

Materials   Discs (1 cm diameter) of stainless steel mesh (9)
Plastic diffusion tubes (3)
Coloured polythene caps (3)
Clear polythene caps (3)

Supplier   Gradko International Ltd
St Martins House
77 Wales Street
Winchester
Hampshire, UK

PRECIPITATION CHEMISTRY

Aim   *To measure the chemical composition of precipitation and dry deposition (bulk precipitation) at ECN sites using a continuously open gauge*

Rationale   Although acid rain and its effects have been of interest for more than 100 years (Smith 1872), most high-quality, systematic studies of acidic deposition have been carried out since the 1970s when observations of freshwater acidification alerted scientists to the need for better and more widespread data.

Subsequent research has highlighted the many impacts which the deposition of atmospherically transported pollutants has on ecosystems (eg Last & Watling 1991). The extent and significance of these impacts on physical, chemical and biotic components of ecosystems are likely to be affected by future changes in emissions of pollutants, and make the measurement of the deposition of atmospherically transported material an important activity for ECN. Deposition from the atmosphere has two components – wet deposition consisting of rainfall or snowfall, and dry deposition consisting of gaseous and particulate material. Dry deposition of pollutants can be calculated for some components as the difference between their concentrations in a continuously open collector and in a collector which is open only during precipitation episodes. It can also be calculated from the concentrations of gaseous and particulate pollutants and of deposition velocities. The use of a continuously open funnel (bulk collector) results in the collection of both dry and wet deposition and, although there are advantages to be gained from separating the two sources of input, it was concluded that only bulk deposition would be measured in ECN on the grounds of the considerable additional costs attaching to the separate measurement of the components.

The main source of data on the geographical distribution of deposition in the UK has been the United Kingdom Precipitation Composition Network, operated until recently by the Warren Spring Laboratory on behalf of the Department of the Environment (Review Group on Acid Rain 1990). The methods used in this Protocol conform with those used at Secondary Sites of the UK Precipitation Composition Monitoring Networks (UKPCMN) described in Devenish (1986). This allows results from ECN sites to be linked with the 32 rural UKPCMN sites at which monitoring started in 1986.

Method   **Equipment**

The bulk collector is constructed to the design described by Hall (1986). It consists of a conical polythene funnel which rests on the neck of a polythene collecting bottle. Two funnel sizes are available, each with a 63° cone; the smaller has a diameter of 115 mm, whilst the larger has a diameter of 152 mm and is likely to be appropriate at all ECN sites. The upper surface of the collector is 1.75 m above the ground. A removable filter of 1 mm mesh teflon prevents coarse debris from falling into the collecting bottle which is surrounded by a jacket of polished steel, from which it is separated by a 25 mm gap. The sample is kept dark and cool by the jacket. The collector has a bird deterrent but this is not always successful and may need to be supplemented by setting up alternative, decoy bird perches some distance from the collector. Details of the required equipment are provided in Appendix II.

Location

Local sources of contamination should be avoided, or their effects minimised, by placing the collector upwind of any such sources. Proximity to vehicle tracks which may become dusty in dry weather, and to animal houses, should be avoided. The collector should be placed in an open location well away from obstructions such as buildings and trees, adjacent to the TSS and as near as is practicable to the soil solution samplers, but it should be borne in mind that regular visits to the collector are likely to cause trampling of vegetation and compaction of the soil. An acceptable alternative location is close to the weather station but each instrument should not cause an obstruction to the other. The collector must be either bolted to a concrete base or secured by guy ropes.

Sampling

The procedure to be used in collecting samples is described in Appendix I, and follows closely the procedures adopted by the Warren Spring Laboratory. Whenever possible the collection bottle should be changed at the same time and on the same day each week; the standard which has been adopted for ECN is Wednesdays at 0900 GMT. The bottle containing the precipitation is removed and is replaced by a clean bottle fitted with a clean filter. The funnel is either cleaned with de-ionised or distilled water and shaken to remove droplets, or is replaced by a funnel which has been cleaned in the laboratory. A funnel containing ice or snow at the time of sample collection must be transferred to the new bottle without having been cleaned or replaced; the ice and snow are thus left to melt into the new bottle when weather conditions permit.

The bottle and its contents are returned to the laboratory, where the volume of precipitation is determined to the nearest 1 ml. To determine the volume, the bottle (without cap) and its contents are weighed and the weight of the dry bottle subtracted to give the weight of contents. Conductivity and pH are measured on unfiltered water according to methods provided in the ECN Initial Water Handling (WH) Protocol which also sets out procedures for filtering the sample. After filtering, the water is analysed for dissolved Na^+, K^+, Ca^{2+}, Mg^{2+}, Fe^{2+}, Al^{3+}, NH_4^+-N, Cl^-, NO_3^--N, SO_4^{2-}-S, PO_4^{3-}-P and alkalinity. In calculating deposition fluxes, precipitation volume should be taken from the nearby automatic weather station or standard raingauge because the bulk precipitation collector may give less accurate volume estimates, eg after snow has fallen.

Labelling

Each water sample is identified uniquely by:
- the ECN Measurement Code (PC),
- the ECN Site ID Code (eg 04 for Moor House),
- the Location Code (eg 01), and
- the collection date ('Sampling Date') (eg 01-Jan-1996).

This information MUST be marked on the sample bottle, so that it can be used to identify the sample through its various analytical stages, and it must accompany the results when transferred to the ECN database. The recording form for this measurement includes codes for factors which potentially affect the chemistry of the sample, such as the presence of insects, dust, or bird droppings in the funnel or bottle, as well as local disturbances such as fires, dust sources, etc.

Washing equipment

Laboratory washing of the funnel, filter and bottle should be carried out with a laboratory cleaning agent, as described for bottle washing in the ECN Initial Water Handling (WH) Protocol.

Quality assurance

QA procedures will follow those adopted for the Acid Deposition Monitoring Network (Heyes, Irwin & Barrett 1985). These include assessment of errors which may be introduced during sampling, sample handling and analysis.

Authors   *J.K. Adamson and J.M. Sykes*

References   **Devenish, M.** 1986. *The United Kingdom Precipitation Composition Monitoring Networks.* Stevenage: Warren Spring Laboratory.

Hall, D.J. 1986. *The precipitation collector for use in the Secondary National Acid Deposition Network.* Stevenage: Warren Spring Laboratory.

PC Protocol

Heyes, C.J., Irwin, J.G. & Barrett, C.F. 1985. Acid deposition monitoring networks in the United Kingdom. In: *Advancements in air pollution monitoring equipment and procedures*, 155–168. Bonn: Federal Ministry of the Interior.

Last, F.T. & Watling, R., eds. 1991. *Acidic deposition, its nature and impacts.* Edinburgh: Royal Society of Edinburgh.

Review Group on Acid Rain. 1990. *Acidic deposition in the United Kingdom 1986–1988.* London: Department of the Environment.

Smith, R.A. 1872. *Air and rain. The beginnings of a chemical climatology.* London: Longmans, Green & Co.

Appendix I   Routine sample collection

Materials
- A clean sample bottle with a screw cap attached. The sample code and the dry weight of the bottle without cap should be marked on the bottle with a water-resistant felt pen
- A wash bottle containing de-ionised or distilled water (DW)
- Forceps
- Two clean funnel filters (in case of one being dropped) in separate, sealed polythene bags
- Two clean funnels (in case of one being dropped) in separate sealed polythene bags
- A clean bottle cap in a sealed polythene bag
- An empty clean polythene bag large enough to contain the funnel assembly during bottle changing
- Field recording form and pencil

Procedure
1. Note any obvious signs of bird droppings, dust, smuts, or any other unusual occurrences or disturbances on the field recording form using appropriate codes, or text if suitable codes do not exist.

2. Detach the funnel assembly from the sample bottle. First unclip the retaining springs, then hold the stem of the funnel and pull it upwards whilst easing the lower end out of the bottle with the other hand. Do not put your fingers inside the top of the bird guard. Be careful not to tip out the funnel insert. Place the funnel and the filter insert inside the polythene bag whilst you deal with the sample bottle.

3. Take the clean cap from the polythene bag and screw it very firmly on to the sample bottle. Be careful to prevent anything entering the bottle and avoid especially the possibility of contamination from your fingers round the neck and rim of the bottle.

4. Remove the sample bottle. The bottle should be raised far enough to allow it to pass sideways between two of the struts supporting the bird guard rings.

5. Insert the clean sample bottle complete with its screw cap. It should be inserted sideways between two of the struts holding the bird guard rings, and then lowered into the inner metal container. Leave the screw cap in position.

6. Remove the filter insert from the funnel with forceps and leave it in the polythene bag.

7. If there is no evidence of significant contamination to the funnel (eg bird droppings), wash the funnel assembly by flushing with

DW. Vigorously shake off the excess water and do not use paper tissues. If there is evidence of significant contamination, or if the funnel has not been washed in the laboratory for six weeks, it should be replaced with a clean funnel. A clean filter insert should be used each week.

8. The screw cap should now be removed from the new bottle and placed in a polythene bag for use on the next visit to the collector.

9. Insert the funnel vertically through the bird guard and position the lower end firmly in the bottle. Attach the retaining clips.

10. Check that the top of the funnel is horizontal.

11. It is **VITAL** to check that the cap is screwed down very firmly on the bottle containing the sample, so that leakage cannot occur during its transport to the laboratory.

Appendix II   Equipment details

Materials

- Complete assembly: aluminium stand, bottle holder, reflector, funnel retainer assembly, polythene catchment funnel assembly, filter and catchment bottle
- Spares: funnel assembly catchment bottle filter

It is recommended that three funnel assemblies, bottles and filters should be available for each site.

Funnel size (usually 152 mm and **not** 115 mm) should be specified.

Supplier

Just Plastics Ltd
Cromwell House
Staffa Road
Leyton
London E10 7P, UK

Sole supplier (as far as is known)

Tel: 0181 558 5238
Fax: 0181 558 3699

SURFACE WATER DISCHARGE

Aim *The continuous recording of stream or river water discharge at sites where the whole catchment area of a natural, perennial stream lies within the ECN site*

Rationale The impact of environmental change is likely to bring about a response in hydrological conditions at a site. The water balance at any location is controlled by climate, vegetation cover and soil properties. Any change in the external climate or in the internal structure of the soil/vegetation system will be reflected in changes in site hydrology. This may involve changes in evaporation, in soil moisture levels, and in the amount of runoff from the site. At sites where snow forms an important component of the precipitation, climate change may have particularly important effects. Monitoring of hydrological variables in ECN may therefore provide a sensitive indicator of environmental change.

Method **Equipment**

Recording of river stage, or level, will use a permanently installed weir or flume whose design will be determined by the conditions at each site but will accord with BS 3680 (BSI 1965). Data are recorded by a Campbell Scientific digital CR10 logger and this should be supplemented where possible by an analogue Ott chart recorder. A dip-flash device will also be installed to facilitate manual readings of stage.

Location

The complete installation comprises an approach channel, a measuring structure, and a downstream channel. The condition of each of these three components affects the overall accuracy of the measurements. In selecting the site, particular attention should be paid to the following:
- the adequacy of length of channel of regular cross-section available;
- the regularity of the velocity distribution over the cross-section of the approach channel;
- the avoidance of a steep channel if possible;
- the effects of any increased water levels upstream due to the structure;
- the impermeability of the ground into which the structure is to be founded;
- the necessity for flood banks to contain the maximum discharge to the channel;
- the stability of the channel downstream of the structure.

Full details are described in BS 3680, Part 4A.

Operation

The logger records stage at ten-second intervals, which the internal program averages and stores over five-minute periods. The logger also calculates an average 15-minute discharge value by taking the mean of the three stage values which it converts to discharge using a rating relationship. Fifteen-minute values of both stage and discharge are stored for quality control of the data and for possible re-calculation if the rating relationship changes. A rating curve will be developed for converting stage to discharge ($m^3 s^{-1}$).

Calibration, by the development of a rating relationship, will be carried out using current meters over a full range of flow conditions and repeated every two years, additionally if the weir approach conditions change. Each site will produce its own calibration protocol. The level of the dip-flash datum will be checked every two years. Data quality control will be carried out by site staff after initial training, according to the procedure detailed in Appendix I.

Data will be downloaded to a PC at fortnightly intervals, *via* a storage module. The procedure for each fortnightly site visit is as follows.

1. Take dip-flash reading.

2. Note time and stage on the old and new Ott chart; remove the old chart and insert the new chart; wind the clock and set the pen to the correct time and stage.

3. Remove the data from the Campbell logger using the storage module.

4. Plug in the key-pad, reset the stage value if necessary and check the logger voltage. If the battery voltage is low (<11.5 volts). check the solar panel (if installed) and its connections to the logger, or if necessary change the battery. Do not remove the old battery without first plugging in a temporary external battery to preserve the internal software.

5. Check the weir approach and crest for debris, sediment or ice accumulation and clear if possible without compromising safety. Check for ice in the stilling well and break up if present. Check that all cables are able to move freely.

It is recommended that a duplicate copy of the logger program should be kept at the site.

Ott charts will be stored for further reference, should this be necessary.

Authors *R.C. Johnson and T.P Burt*

Reference **British Standards Institution.** 1965. *BS 3680. Methods of measurement of liquid flow in open channels. Part 4. Weirs and flumes. 4A. Thin plate weirs and venturi flumes.* London: BSI.

Appendix I   Quality control of surface water discharge data

Procedure

1. Check that the first and last data points agree with times of data removal from logger.

2. Check manual (dip-flash) and automatic stage readings at start and finish.

3. Compare 12 noon stage values from both sensors on days when steady flow.

4. Note any missing data or times when weir could have been frozen.

5. Make notes on results of all checks, changes and maintenance (eg dredging) carried out for inclusion in the ECN database.

SURFACE WATER CHEMISTRY AND QUALITY

Aim *Collection of weekly stream/river water samples for analysis of major cations and anions*

Rationale The possible causes and consequences of changes in climate, land use and industrial, urban and agricultural pollution can be expected to be indicated by changes in the physical and chemical composition of water in rivers and streams. These are determined by both biogeochemical processes and by changes in terrestrial or atmospheric inputs. In addition to the major ions (eg Na^+, K^+, Ca^{2+}, Mg^{2+}, Cl^-, SO_4^{2-}-S, alkalinity) which give a measure of the fundamental chemical composition of the water, other variables (eg NH_4^+-N, NO_3^--N, PO_4^{3-}-P) are major plant nutrients, whilst yet others (eg Al_3^+) have chemical states which are pH dependent and are of particular interest where water bodies are undergoing acidification. By measuring concentrations of these ions at frequent and regular intervals, together with water flow, it is possible to calculate loads and fluxes of ions which may be of interest in relation to nutrient budgets.

Method ## Location

Dip samples of flowing river water should be taken from at least one representative site above the weir, the location of which should be recorded and used for sampling on each occasion.

Sampling

Weekly samples will be collected on the same day each week, preferably on Wednesdays so as to synchronise with other ECN water collections (see PC and SS Protocols), within one hour of a time, to be specified for each site, which will depend on site accessibility. One 250 ml sample of river water will be taken from flowing water, using the same location on each occasion. The collecting bottle will be rinsed with river water, shaken vigorously with the stopper in place, and then the rinse-water will be discarded. The bottle is filled, leaving no air space, by reaching upstream. Stage height will be recorded at the time of sampling to the nearest 2 mm. Conductivity and pH will be measured on unfiltered water according to methods described in the Initial Water Handling (WH) Protocol which also describes the filtering and subsequent treatment of the sample before further analysis.

After filtering, the water is analysed for Na^+, K^+, Ca^{2+}, Mg^{2+}, Fe^{2+}, Al^{3+}, NH_4^+-N, Cl^-, NO_3^--N, SO_4^{2-}-S, PO_4^{3-}-P, alkalinity and dissolved organic carbon.

Washing equipment

Procedures for bottle washing are described in the Initial Water Handling (WH) Protocol.

Labelling

Each water sample is identified uniquely by:
- the ECN Measurement Code (WC),
- the ECN Site ID Code (eg 04 for Moor House),
- the Location Code (eg 01), and
- the collection date ('Sampling Date') (eg 01-Jan-1996).

This information MUST be marked on the sample bottle, so that it can be used to identify the sample through its various analytical stages, and it must accompany the results when transferred to the ECN database.

Continuous monitoring

It is intended that continuous measurements of stream or river water temperature, conductivity, pH and turbidity should be carried out at suitable ECN sites. Instruments for this purpose are being tested and an ECN specification will be drawn up when satisfactory sensors have been identified.

Authors *R.C. Johnson and T.P. Burt*

S Protocol	**SOILS**

Aim	*To classify the soils at ECN sites, to characterise and quantify their physical, chemical and mineralogical properties and to quantify temporal changes in those properties*

Rationale

Patterns of soil and vegetation are closely interlinked in areas of natural and semi-natural vegetation, the pattern of vegetation and the distribution of individual plant species both reflecting and influencing the complex interaction of the soil's chemical, physical and biological properties. Changes in soil characteristics may result in changes in vegetation, and *vice versa*. The link between soils and surface vegetation patterns is largely broken where land is under intensive agricultural management but the inherent soil properties are nevertheless major factors influencing the choice of crops and the type of land management.

Soils also exert a strong influence on surface water chemistry and flow regimes, on gas exchange between the atmosphere and the earth's surface, and on the fate of deposited pollutants. Changes in soils take place naturally over time but can also be driven by alterations in pollution climate and land management.

The soil map compiled at the start of the programme and the associated characterisation of the soils comprising the various map units provide the baseline against which changes over time can be assessed. A complete characterisation of a soil requires information on physical, chemical, mineralogical and biological attributes which change at different rates in response to changes in, for example, climate, pollution or land use. Thus, soil chemistry tends to respond more rapidly than do physical characteristics such as texture and structure. Exchangeable soil chemistry, comprising the fraction most readily available for plant uptake and loss to drainage waters, responds more readily than total soil chemistry. Similarly, various aspects of soil biological properties change at different rates. Sampling intervals and their associated analyses have been planned to take account of these differences.

Method

Soil survey and classification (SB)

An extensive survey of the whole of the ECN site is carried out, using a soil auger and spade, to provide maps at a scale of 1:10 000 for sites up to 50 km^2 and 1:25 000 for sites over 50 km^2; map units will be identified to soil series level, or complexes where necessary (Avery 1980), but will also be classified according to the systems of the Food and Agriculture organisation and of the US Department of Agriculture (FAO-UNESCO 1974; FAO 1986; USDA 1975). Where site soil maps already exist, these will be evaluated and the necessity for additional survey work will be assessed by the appropriate Soil Survey organisation.

An intensive survey will be carried out on a 300 m x 300 m area, with the TSS at its centre, using auger borings at 50 m grid intersections over the whole 9 ha area and at 25 m grid intersections within the TSS (Figure 6). Where possible, the TSS is oriented towards north (see LM Protocol).

Method **Soil characterisation and assessment of change (SF and SC)**

Sampling frequency and location

Soil sampling will be carried out at the beginning of the programme. The soils are subsequently re-sampled at five-yearly and 20-yearly intervals, using different sets of determinands over the two periods. Both periodic samples are replicated in six blocks of which four are located adjacent to the sides of the TSS and two are close to its centre. The side blocks may be square and located immediately outside the TSS as suggested in Figure 7, or linear and located immediately inside the TSS as suggested in Figure 8; the choice will depend on the particular conditions attaching to the TSS at a site. The arrangement suggested in Figure 7 can be used

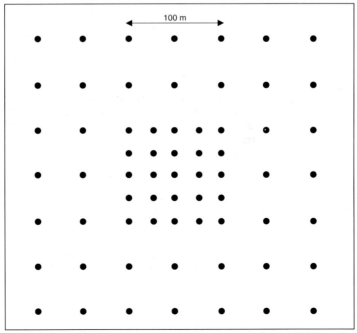

Figure 6. Distribution of auger bores for intensive soil survey of TSS and surrounding area

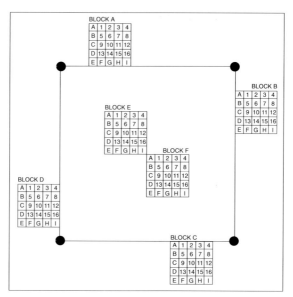

Figure 7. Suggested layout of soil sampling blocks in the TSS

Cells A–I in blocks A–F are used for 20-year sampling, each cell being 5 m x 5 m. On each sampling occasion one cell is selected at random from each block, ie a total of six samples on each sampling occasion

Cells 1–16 in blocks A–F are used for five-year sampling, each cell being 5 m x 5 m. On each sampling occasion one subcell (1 m x 1 m) from each cell is selected at random from each block. The 16 samples from each block are bulked to give a total of six samples on each sampling occasion

BLOCK A

		A	B	C	D	E	F	G	H	I	J	K	L	M	N	O	P		
		1	2	3	4	5	6	7	8	9	10	11	12	13	14	15	16		
P	16																	16	P
O	15																	15	O
N	14			BLOCK E														14	N
M	13			A	1	2	3	4										13	M
L	12			B	5	6	7	8										12	L
K	11			C	9	10	11	12										11	K
J	10			D	13	14	15	16										10	J
I	9			E	F	G	H	I	BLOCK F									9	I
H	8								A	1	2	3	4					8	H
G	7								B	5	6	7	8					7	G
F	6								C	9	10	11	12					6	F
E	5								D	13	14	15	16					5	E
D	4								E	F	G	H	I					4	D
C	3																	3	C
B	2																	2	B
A	1																	1	A
		A	B	C	D	E	F	G	H	I	J	K	L	M	N	O	P		
		1	2	3	4	5	6	7	8	9	10	11	12	13	14	15	16		

BLOCK D (left side) · BLOCK B (right side) · BLOCK C (bottom)

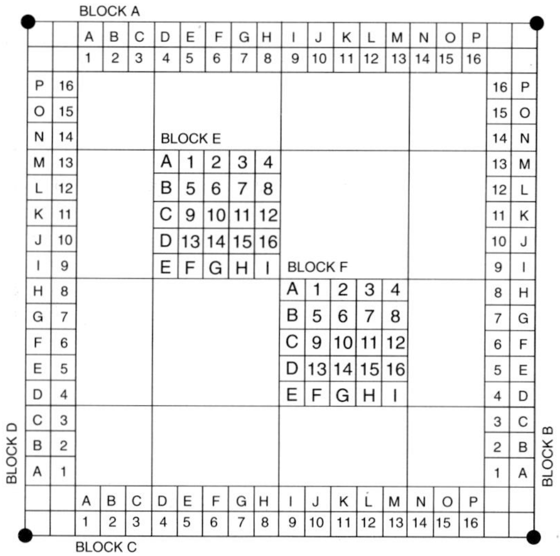

Figure 8. Alternative layout of soil sampling blocks in the TSS

Cells A–P in blocks A–D and cells A–I in blocks E and F are used for 20-year sampling, each cell being 5 m x 5 m. On each sampling occasion one cell is selected at random from each side block (blocks A–D) and from each inside block (blocks E and F), ie a total of six samples on each sampling occasion

Cells 1–16 in blocks A–F are used for five-year sampling, each cell being 5 m x 5 m. On each sampling occasion one subcell (1 m x 1 m) from each cell is selected at random from each block. The 16 samples from each block are bulked to give a total of six samples on each sampling occasion

where the TSS has fragile vegetation and where it is important to avoid damage; the arrangement suggested in Figure 8 may be used where the availability of space outside the TSS is very limited. Where neither design is thought suitable for local circumstances any proposed alternative must be approved by the ECN Central Co-ordination Unit.

Each sampling block is referenced alphabetically as shown in Figures 7 and 8, the most northerly block being designated 'A' and the others logically in a clockwise direction. The corners of each block should be marked permanently with wooden posts or stone blocks and recorded on a map. Each sampling block is itself divided into 5 m x 5 m cells, of which some are used for five-yearly sampling and others for 20-yearly sampling, as shown in Figures 7 and 8. Each cell used for five-yearly sampling is assigned a number from 1 to 16; each cell used for 20-yearly sampling is assigned an identifying letter from A to I (or A–P in the case of linear blocks), as shown in Figures 7 and 8. The corners of the cells should be marked temporarily by canes for convenience in locating current sampling positions. The location of each cell used on any sampling occasion is recorded on a plan.

Sampling – initial and five-yearly (SF)

Five-yearly sampling is carried out in each of the 16, numbered, 5 m x 5 m cells in each of the six blocks. Each 5 m x 5 m cell is subdivided into 25 subcells of 1 m x 1 m which are numbered as shown in Figure 9, with the cell orientated with the TSS. On each sampling occasion only one subcell is randomly selected from each 5 m x 5 m cell, giving a total of 16 sampling sites for each

1	2	3	4	5
6	7	8	9	10
11	12	13	14	15
16	17	18	19	20
21	22	23	24	25

N

Figure 9. Numbering the 25 1 m x 1 m subcells for random selection of five-yearly samples

block at each five-yearly sampling. For convenience it is necessary to locate only each subcell which is to be used for sampling; the centre of the subcell should be marked with a cane. Soil sampling will be carried out within a small area around this central point. In subsequent five-yearly samples, different 1 m x 1 m subcells are selected, and thus no subcell is sampled more than once during the ECN project. The location of the subcell to be used for sampling in each block and cell will be recorded.

Sampling should be carried out using a gouge auger of a diameter suitable to provide sufficient bulked sample (about 3 kg for mineral soil and at least 6 kg for organic soil, depending on water content) for the soil type being sampled. For slightly stony soils, a narrower auger may be preferred. For some stony soils or soils with very thin horizons, it will be necessary to excavate a small inspection pit with a spade and sample from the exposed face.

Two sets of soil samples will be taken to a maximum depth of 30 cm from each sampled subcell. One set is based on depths: 0–5 cm, 5–10 cm, 10–20 cm, and 20–30 cm. The other set corresponds to horizons within the top 30 cm. It should be noted that soil horizons within the upper 30 cm can be difficult to identify clearly when using a soil auger. Horizon boundaries often merge over a depth of several centimetres and this gradation from one horizon to another may be thicker than the actual horizons. In practice, a soil layer should be designated a horizon when organic matter content or colour show a change.

The 16 subsamples by horizon and 16 subsamples by depth from each block should be bulked for each block. In some soil types with thin horizons, it may be necessary to take several auger samples within each subcell to ensure that sufficient soil is obtained.

Sampling – initial and 20-yearly (SC)

Profile sampling and description require excavation of the ground to expose a vertical section of soil suitable for description. Sampling and description will use standard methods (Hodgson 1974) and will be from six pits, each located in an alphabetically labelled 5 m x 5 m cell (Figures 7 & 8) chosen at random from each block. A modified version of the Von Post scale (Avery 1980), together with the vegetation composition scheme of Troels-Smith (1955), will be used for describing the decomposition state of peats.

Profile description and sampling can be carried out at most times of the year, but preferably when the soil is at, or fairly close to, field capacity. Orientation of the profile face should be in the direction of greatest sunlight if possible but space limitations in the area of the sampling cell may preclude this. During excavation topsoil should be kept separate from subsoil and care should be taken to avoid contaminating the surface of adjoining sample cells when using the design suggested in Figure 8. It is advisable to lay a polythene sheet upon which to place the spoil; this allows a tidier re-instatement. The excavation should expose a soil profile face of suitable size and of sufficient dimensions to allow easy working. A soil profile should be exposed to 1.2 m if possible. **It should be noted that this is the maximum depth allowed by the Health and Safety Executive before shoring of the excavation sides is considered necessary.**

Using a trowel, knife or similar instrument, the profile face should be picked back to expose the soil structure from the surface downwards. This will allow identification of the soil horizons from their colour, texture and structural development and, by using a 10% solution of hydrochloric acid, the presence of excess calcium carbonate. The depth of each horizon, measured from the soil surface, should be recorded together with the location of the site, the soil surface description and the description of each horizon, following the procedures presented in the Soil Survey field handbook (Hodgson 1974). It may also be useful to record any other observations of interest which are not specified in the handbook. The description and characteristics of the site and of each horizon are recorded either by tape-recorder or in a notebook for subsequent transfer to computer and hard copy.

Samples should be collected from each soil horizon recognised in the description to about 1 m depth (or less if rock is encountered) and by standard depths of 0–5 cm, 5–10 cm, 10–20 cm, 20–40 cm, 40–60 cm, 60–80 cm, 80–100 cm, and 100–120 cm. The positions of the depth bands relative to the horizons should be recorded. Each horizon should be sampled from its full depth beginning with the lowest (deepest) so as to avoid contamination of other, higher horizons; 3–4 kg of soil should be collected from each horizon and from each depth band in each soil pit. Samples should be collected into plastic bags with the sample bag then placed within a second plastic bag and a label placed between the two bags. All samples should be stored at 4–6°C prior to drying and analysis. Samples should be stored for as short a time as possible to minimise the effect of biological activity within the cores.

In addition, core samples will be taken in triplicate from each horizon, using the methods given in Hodgson (1976), for the measurement of soil water release characteristics and of bulk density. It may be possible to take cores from depths which correspond with the fixed depth samples.

The location of each sampled cell should be marked, preferably both on the ground and on a plan.

Please note that *both* SF and SC are carried out in years 20, 40, etc.

Sample handling and storage

Soil samples should be stored in clearly marked polythene bags and kept in the dark at 4–6°C prior to drying and analysis. Grinding facilities should satisfy criteria set by the soil analysis subgroup. All bulk samples will be air-dried and sieved at 2 mm prior to analysis, except where a sample is required for analysis in the moist state, in which case a moist subsample will be taken by an accepted method. Chemical analyses are carried out on the fine earth unless otherwise stated.

Particle size analysis

Particle size analysis should be carried out on each horizon of the profile samples collected in the initial sampling (see SC above) from the six profile pits and using the pipette/peroxidised soil method as defined in the Soil Survey laboratory handbook (Avery & Bascomb 1982) and which complies with international particle size analysis standards.

Soil mineralogy

Analyses should be on clay (<2 µm), silt ($2–63$ µm), and coarse silt and sand (63 µm$–2$ mm) fractions from each of the bulked profile horizon samples taken from the initial sampling (see SC above), using a combination of x-ray diffraction and optical microscopy techniques. Analysis of the heavy mineral fraction (SG >2.65) should also be carried out. Only clay mineral analysis will be repeated at 20-year intervals.

Soil chemistry

Each bulked horizon and depth band sample from the five-yearly (*) core samples and each horizon and depth band from the 20-yearly (†) profile samples will be analysed for:

*†	Moisture	on <2 mm soil oven dried overnight at 105°C
*†	pH	on field moist and air dry samples, on 1: 2.5 extracts in water and 0.01 M calcium chloride

Exchangeable

*†	acidity	0.5 M barium acetate pH 7
*†	sodium	1 M ammonium acetate pH 7 and unbuffered
*†	potassium	1 M ammonium acetate pH 7 and unbuffered
*†	calcium	1 M ammonium acetate pH 7 and unbuffered
*†	magnesium	1 M ammonium acetate pH 7 and unbuffered
*†	manganese	1 M ammonium acetate pH 7 and unbuffered
*†	aluminium	1 M potassium chloride pH 7

Total

*†	nitrogen	Kjeldahl digestion
*†	phosphorus	NaOH fusion (Smith & Bain 1982)
*†	sulphur	(to be decided)
*†	organic carbon	dichromate digest
*†	inorganic carbonate	manometric
†	lead	aqua regia
†	zinc	aqua regia
†	cadmium	aqua regia
†	copper	aqua regia
†	mercury	nitric acid/sulphuric acid digest
†	cobalt	aqua regia
†	molybdenum	aqua regia
†	arsenic	$Mg(NO_3)_2$ ashing, sulphuric acid/citrate extraction
†	chromium	aqua regia
†	nickel	aqua regia

Extractable

†	iron	ammonium oxalate pH 3
†	aluminium	ammonium oxalate pH 3
†	phosphorus	0.5 M sodium hydrogen carbonate

Bulk density and water release characteristics

These should be determined for each major soil horizon sampled as part of the 20-year sampling (see SC above). The method will be the Soil Survey suction table and pressure system (in triplicate) procedures in Hodgson (1976). Bulk density measurements will be carried out in conjunction with water release measurements.

Archiving of soil samples (SA)

Soil preparation before storage

The field samples must be allocated a unique ECN site, sample number and date for reference purposes. The sample is spread out thinly on a tray and dried at 25°C. Drying may take up to five days for a mineral soil and up to 30 days for a highly organic soil or peat, depending on its water content. The sample is then crushed or rolled to break up any aggregates and sieved through a 2 mm sieve. Highly organic samples are ground in a cross-beater mill. The >2 mm material which is retained after sieving generally consists of stones, roots or other organic matter and is discarded. Before any form of analysis, the <2 mm soil is homogenised by a mechanical sample divider (eg Fritsch Laborrette 27) which produces eight uniform subsamples. This ensures that the analysed samples and those stored for future work are exactly similar.

Storage

The sieved, homogenised soils are stored in high-density plastic bottles or food-quality polycarbonate bags. Samples stored in bottles should have the sample number shown on the lid and on the container itself. Samples stored in bags should be placed inside a second, similar bag; the outer bag should carry the sample number, and the sample number should also be written on a tag placed between the inner and outer bag. Subsamples of at least 500 g, or 100 g for highly organic soils, are stored in glass jars with sealed lids, and labelled.

Authors   *M. Hornung, G.R Beard, J.M. Sykes and M.J. Wilson*

References   **Avery, B.** 1980. *Soil classification for England and Wales (higher categories).* (Technical monograph no.14.) Harpenden: Soil Survey.

Avery, B.W. & Bascomb, C.L. 1982. *Soil Survey laboratory methods.* (Technical monograph no.6.) Harpenden: Soil Survey.

FAO-UNESCO. 1974. *Soil map of the world 1:5,000,000 vol. 1: legend.* Paris: UNESCO.

Food and Agriculture Organisation. 1986. *Soil map of the world 1:5,000,000: revised legend.* (World soil resources reports no.58.) Rome: FAO.

Hodgson, J.M. 1974. *Soil Survey field handbook.* Harpenden: Soil Survey.

Smith, B.F.L. & Bain,D.C. 1982. A sodium hydroxide fusion method for the determination of total phosphate in soils. *Soil Science and Plant Analysis,* **13**, 185–190.

Troels-Smith, J. 1955. *Karakterisering af løese jordarter.* Copenhagen: Geological Survey of Denmark.

United States Department of Agriculture. 1975. *United States Soil Conservation Service soil taxonomy: a basic system of soil classification for making and interpreting soil surveys.* (Agriculture handbook no. 436.) Washington, DC: US Government Printing Office.

SOIL SOLUTION CHEMISTRY (SS)

Aim   *To monitor changes in the chemical composition of soil solution using suction lysimeters*

Rationale   Soil solution chemistry is likely to be affected by physical and chemical changes in the environment and itself to have important effects on ecosytem processes; for example, soil temperature and moisture status strongly influence the microbial activity in soil which controls nutrient release into the soil solution (Swift, Heal & Anderson 1979). In turn, soil solution nutrient status has a major influence on plant productivity and thus on animal productivity.

The commonly used non-destructive methods of soil solution sampling can be divided into zero tension methods and suction methods. Zero tension methods use a volume of soil contained in such a way that water percolating through the soil drains into a collecting vessel. These methods are suitable for leaching studies, where the objective is to quantify nutrients being lost from the soil (Addiscott 1990). Suction methods involve the insertion into the soil of a porous-walled sampler which is evacuated so that water is drawn into it. Suction samplers collect water whether or not it is percolating through the soil and were termed 'artificial roots' by early researchers (Briggs & McCall 1904). For the purposes of ECN, it has been thought most appropriate to sample water which equates approximately with that which is available to plants, and accordingly suction sampling is being used.

Ceramic suction samplers have been criticised because the chemical composition of the soil solution can be modified by contact with the ceramic material (Raulund-Rasmussen 1989), and for this reason samplers manufactured by Prenart from PTFE and quartz are specified. The small size of Prenart samplers makes them suitable for installation in the wide variety of soils encountered at ECN sites; larger and more intrusive types of sampler would result in hydrological disruption and aeration of peat soils, and would be impossible to install at the base of shallow soils, in stony soils and in soils containing tree roots.

Suction samplers are not without operational problems, a number of which were identified by Hanson and Harris (1975), who also suggested procedures to overcome them. While factors such as soil texture can influence the characteristics of the soil solution collected at a site, this is likely to be of minor importance in an environmental change study because the main emphasis here is on temporal variation rather than on between-site variation. 'Plugging' of the sampler pores by soil particles may cause the performance of samplers at some sites to change through time; in

75

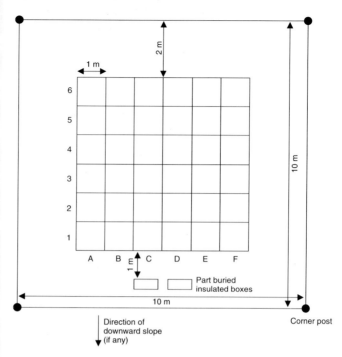

Figure 10. Layout of soil solution sampling area

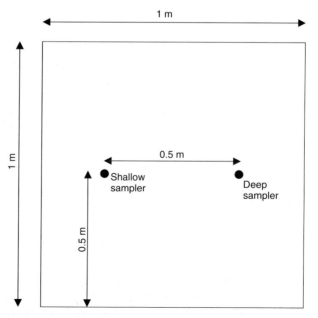

Figure 11. Location of samplers within a 1 m x 1 m cell

order to monitor this possibility the volume of water collected and the residual vacuum must be recorded. A procedure for sampler replacement is included in this Protocol.

Method

Equipment

Prenart 'super quartz' soil water samplers are manufactured from PTFE and silica flour. They are cylindrical, 21 mm in diameter and 95 mm in length, conical at one end and with a tube attachment at the other. The tubing links the sampler to a 1 litre glass collecting bottle with a Prenart screw cap. The collecting bottles are placed in an insulated box to protect the samples from extremes of temperature and are evacuated using a portable pump. Details of the equipment required are provided in Appendix I.

Location

Samplers will be located within a 6 m x 6 m plot which is itself located within a 10 m x 10 m plot on the edge of, but outside, the TSS. On sloping sites the plot will ideally be located on the downslope edge of the TSS and definitely not on the upslope edge, so as to avoid any debris from soil disturbance and trampling being washed on to the TSS. Although no permanent location markers are used within the 10 m x 10 m plot, it should be envisaged as being divided into 1 m x 1 m cells, as shown in Figure 10.

Six samplers will be installed in the A horizon and six others at the base of the B horizon. The conical base of the sampler can be allowed to penetrate the horizon below the one being sampled, but the porous section of the sampler should not do so. Where these horizons do not exist (eg in peat soils) depths of 10 cm and 50 cm will be used. In shallow soils where these horizons do not exist, the upper samplers will be installed at 10 cm and the lower samplers at the base of the soil. In soils less than 20 cm in depth, only one set of six samplers will be installed, at a depth of 10 cm. These soil depths are depths below and perpendicular to the surface, and refer to the position of the mid-point along the length of the sampler.

The procedure for selecting cells to receive samplers and also for the rolling programme of sampler replacement is described in Appendix II. Each selected cell will have both a deep and a shallow sampler located according to Figure 11. Samplers should be identified by a three-character code, where the first character is the letter co-ordinate for the cell (as given in Figure 10), the second character is the number co-ordinate for the cell and the third character is either 'S' for shallow or 'D' for deep. The installation procedure is described in Appendix III. Trampling within the plot during sampling should be minimised, though animals grazing the TSS should also have access to the soil solution sampling plot.

Sampling

The detailed procedure for collecting samples is provided in Appendix IV. Samplers will be emptied and water volumes recorded on the same day each fortnight, synchronised to coincide with Wednesday, 3 January 1996. One week after sample collection the samplers should be evacuated to 0.5 bar; thus the water sample accumulates over only the second week of the fortnightly period. In some clay soils this partial vacuum will be insufficient to extract a sufficient quantity of soil solution to allow chemical analysis; if this is found to be the case after several weeks of wet weather, then a vacuum of 0.7 bar should be applied following agreement with the ECN Central Co-ordination Unit; subsequent sampling at such sites should always be at 0.7 bar.

Ideally soil solution should be collected throughout the year but in dry periods the volume of extracted water will decrease and eventually no further water will be extractable. Some sites will be particularly prone to this condition and, even in an average summer, there may be a substantial period when no water is extracted. There is little point in continuing fortnightly evacuation during such periods but it is important to ensure that sampling is resumed as soon as the soil becomes sufficiently wet because early samples are likely to have particularly high nutrient concentrations. Close attention should therefore be paid to meteorological data from the site and the vacuum appropriate to the site should be applied periodically to determine whether water can be extracted. The dates of these vacuum checks, and their results, should be included with the normally formatted soil solution data forwarded to the ECN database. The standard fortnightly evacuation routine should be operated regularly at all sites from the beginning of October to the end of May, irrespective of soil moisture status.

At some sites, particularly in late spring and in autumn, the volume of water collected may be very small and it may be necessary to discard very small samples, or combine the six samples for analysis. The following rule will be used.

- If the volume collected from any individual sampler is less than 10 ml, it should be discarded.

- If, for a particular soil depth, the volume collected by three or more samplers exceeds 60 ml, samples will be treated individually.

- Otherwise, all samples from that depth will be combined in the laboratory to make a single, combined sample.

Where samples are to be combined, the single combined sample from the shallow soil water samplers will be coded 'XXS' and from the deep soil water samplers 'XXD'.

Conductivity and pH are measured on unfiltered water according to methods in the Initial Water Handling (WH) Protocol, which also deals with filtering of the sample. After filtering, the water is analysed for dissolved Na^+, K^+, Ca^{2+}, Mg^{2+}, Fe^{2+}, Al^{3+}, NH_4^+-N, Cl^-, NO_3^--N, SO_4^{2-}-S, PO_4^{3-}-P, alkalinity and dissolved organic carbon.

Labelling

Each water sample is identified uniquely by:
- the ECN Measurement Code (SS),
- the ECN Site ID Code (eg 04 for Moor House),
- the Location Code (eg 01), and
- the collection date ('Sampling Date') (eg 01-Jan-1996).

This information MUST be marked on the sample bottle, so that it can be used to identify the sample through its various analytical stages, and it must accompany the results when transferred to the ECN database.

Washing equipment

Bottle washing is described in the WH Protocol. The glass collecting bottles attached to the samplers must be washed in the laboratory at six-monthly intervals.

Safety note

Although the suction samplers do not use a high vacuum and the bottles are of toughened glass, there still exists a small risk of injury from bottle implosion. Appropriate safety precautions should therefore be taken, including the checking of bottles for cracks in the glass if, for example, the bottles have been dropped or if it is possible that the contents have been frozen.

Author   *J.K. Adamson*

References   **Addiscott, T.M.** 1990. Measurement of nitrate leaching: a review of methods. In: *Nitrates, agriculture, eau,* edited by R Calvert, 157–168. Paris: INRA.

Briggs, L.J. & McCall, A.G. 1904. An artificial root for inducing capillary movement in soil moisture. *Science,* **20**, 566–569.

SS Protocol

Hanson, E.A. & Harris, A.R. 1975. Validity of soil-water samples collected with porous ceramic cups. *Soil Science Society of America Proceedings*, **39**, 528–536.

Raulund-Rasmussen, K. 1989. Aluminium contamination and other changes of acid soil solution by means of porcelain suction cups. *Journal of Soil Science*, **40**, 95–101.

Swift, M.J., Heal, O.W. & Anderson, J.M. 1979. *Decomposition in terrestrial ecosystems.* (Studies in Ecology, vol. 5.) Oxford: Blackwell Scientific.

SS Protocol

Appendix I   Equipment details (including spares)

Materials

- Prenart 'super quartz' soil water samplers, each with 10 m tubing attached (14)
- Pyrex 1 litre collecting bottles with Prenart screw caps (14)
- Battery-driven 12 volt vacuum pump with charger

Supplier

Prenart Equipment ApS
Buen 14
K-2000 Frederiksberg   *Tel: 01045 3174 1664*
DENMARK   *Fax: 01045 4280 3607*

It should be specified clearly that 10 m of tubing is required on each sampler.

Appendix II   Sampler location and replacement

The sampling arrangement uses some of the properties of the Latin square design. It has the advantages of simplicity, together with the desirable property of an element of randomness for future analysis.

The 6 m x 6 m area into which the soil solution samplers are placed is subdivided into 1 m x 1 m cells. In Year 1 the samplers are placed at two depths in the U cells.

After the first five years samplers in three of the cells, chosen at random, are abandoned and replaced by other samplers in three of the V cells in either the rows or columns. Following replacement either each row or each column will contain a cell with samplers.

After the next five years the remaining samplers in U cells are abandoned and replaced by samplers in the remaining V cells. Each row and column now contains a cell with a sampler.

This procedure is continued in succeeding five-year periods until all 36 cells have been exhausted. The cells in use are underlined in the following diagrams.

Author   *J.N.R. Jeffers*

Year 1	6	W	Y	Z	U̲	V	X
	5	X	V	U̲	W	Y	Z
	4	V	Z	W	X	U̲	Y
	3	U̲	W	Y	Z	X	V
	2	Z	X	V	Y	W	U̲
	1	Y	U̲	X	V	Z	W
		A	B	C	D	E	F

Year 6

	A	B	C	D	E	F
6	W	Y	Z	U	<u>V</u>	X
5	X	V	<u>U</u>	W	Y	Z
4	V	Z	W	X	<u>U</u>	Y
3	U	W	Y	Z	X	<u>V</u>
2	Z	X	V	Y	W	<u>U</u>
1	Y	U	X	<u>V</u>	Z	W

Year 11

	A	B	C	D	E	F
6	W	Y	Z	U	<u>V</u>	X
5	X	<u>V</u>	U	W	Y	Z
4	<u>V</u>	Z	W	X	U	Y
3	U	W	Y	Z	X	<u>V</u>
2	Z	X	<u>V</u>	Y	W	U
1	Y	U	X	<u>V</u>	Z	W

Year 16

	A	B	C	D	E	F
6	<u>W</u>	Y	Z	U	V	X
5	X	<u>V</u>	U	W	Y	Z
4	<u>V</u>	Z	W	X	U	Y
3	U	<u>W</u>	Y	Z	X	V
2	Z	X	<u>V</u>	Y	W	U
1	Y	U	X	V	Z	<u>W</u>

Appendix III   Installation of equipment

Materials

- Water samplers, each with 10 m of tubing attached (12)
- Collecting bottles with caps (12)
- Joiner's auger, 0.75 inch or 2 cm diameter with extended handle for deeper samplers
- Plastic pipe, maximum diameter 2 cm
- Distilled or deionised water
- Plastic beaker (3 of approximately 1 litre)
- Thin orange polypropylene string
- Insulated picnic boxes (2) each to contain six collecting bottles with holes drilled according to para 11 below
- Spade
- Soil sieve (2 mm)
- Vacuum pump
- Canes (24) for temporary marking of plot
- Measuring tape

Procedure

1. New Prenart soil water samplers are already rinsed in HCl and deionised water, ready for installation.

2. Sink the insulated boxes into the soil as indicated in Figure 10 so that only the lids and 5 cm of the walls protrude above the ground surface. While doing this the profile can be examined and soil obtained for para 3 below.

3. Sieve soil from all the horizons encountered. The soil should not be dried prior to sieving and it will usually be necessary to force the soil through the sieve, for instance with a rubber bung.

4. Mix, in the plastic beaker, a thin slurry of water and sieved soil from the horizon into which the sampler is to be placed.

5. Place the soil water sampler in the slurry and apply 0.5 bar vacuum for 10–15 minutes. This ensures that the biggest pores in the sampler are filled with fine soil and that there is a tight capillary contact with the soil.

6. Make a hole with the auger at an angle of 60° to the soil surface until the appropriate horizon/depth is reached. This angle ensures that the soil over the sampler is undisturbed.

7. Mix a thicker slurry of water and sieved soil from the horizon into which the sampler is to be placed and pour it into the hole using the funnel. The tubing attached to the funnel should extend to the bottom of the hole.

8. Tie the string to the top fitting of the sampler leaving enough string so that a loop extends approximately 5 cm above the

ground surface. This serves to mark the location of the sampler and aid its removal should this be necessary.

9. Put the tube from the sampler and the marker string through the plastic pipe and push the sampler down the hole with the pipe, using the string to hold the sampler tightly against the bottom of the pipe.

10. Back-fill the hole with thick slurry made with sieved soil from the appropriate horizons.

11. Tubing from each sampler should be routed directly to the downhill boundary of the 1 m x 1 m cell in which it is located, and it should then follow cell boundaries to an insulated box. In most cases the full 10 m of tubing will not be required to link sampler and bottle. However, the excess tubing should not be cut off but rather coiled under a turf adjacent to the insulated boxes. To minimise the impact of frost and animals, the tubing should be placed in the bottom of a 10 cm deep slit cut in the ground with the spade. The holes in the boxes should allow the tubing to enter without it having to emerge above the ground. These holes should be a tight fit around the tubing to prevent soil or water entering the boxes. In soils which frequently have high water tables there is a danger of the boxes floating. This can be obviated by enlarging the hole and placing at the bottom a plank of wood approximately 150 cm x 30 cm x 5 cm to which the boxes are tied. The plank and boxes are held in place by back-filling the soil.

12. Connect the tubing to the collecting bottles and evacuate the bottles to 0.5 bar using the pump.

13. The depth of the samplers and the layout of the tubing should be carefully recorded for future reference.

Some site conditions may require different installation procedures and experimentation with a limited number of samplers to find the most appropriate procedures is advisable. For peat soils the slurry will not be required; the fibrous peat surface should be penetrated with the auger and the sampler should then be pushed through undisturbed peat to the appropriate depth; if necessary, the hole above the sampler should be back-filled by poking solid peat down with the installation pipe. Where the samplers are installed close to the soil surface because the A horizon is shallow, it may be most appropriate to angle the samplers so they are almost horizontal. Where the string is likely to be damaged, for instance by grazing animals, it should still be installed to allow removal of the samplers but it should be protected by leaving the upper end below the ground surface with the tubing.

Appendix IV   Routine sample collection

Materials
- 250 ml polypropylene bottles, pre-marked with sampler codes (12)
- Vacuum pump (fully charged)
- Recording sheet and pencil

Procedure

1. Connect the pump to a collecting bottle and open the pinch clip. Ensure that the tubing walls have separated and record any residual vacuum.

2. Remove the collecting bottle and record the quantity of water it contains, using the calibrations on the bottle.

3. Fill the polypropylene bottle labelled with the corresponding sampler code and secure its cap.

4. Discard any water remaining in the collecting bottle away from the sampling square, ensuring that as much water as possible is removed from the bottle.

5. Re-connect the collecting bottle, in preparation for evacuation one week later.

6. One week after collecting the sample for analysis, attach the vacuum pump and establish the required vacuum (see under **Sampling**). The Database Manager should be informed of the vacuum being used. The vacuum should be read with the pump switched off because the gauge gives a false value when the pump is operating.

VEGETATION

Aim   *To monitor change in semi-natural vegetation, permanent grass and cereals*

Rationale   Semi-natural and managed vegetation is often very sensitive to the main drivers of environmental change, ie climate, pollutants and land use practices. Its monitoring requires as a starting point an accurate and comprehensive description of the extent and character of the vegetation cover of each ECN site. Referring this variation in vegetation to a descriptive scheme that is applicable at least to all network sites, and preferably to a wider geographical area, is important for comparing changes across ECN. The monitoring methods must be sufficiently sensitive to detect responses in any element of the vegetation, but must also set repeatable standards for both present and future recording and analysis. It has been concluded therefore that it is better to use an objective method which records presence and absence of plant species rather than to attempt difficult and subjective assessment, such as cover estimation. Further, it has been thought more efficient statistically to use a relatively large number of small plots rather than a small number of larger plots.

The basic unit for recording and storing lists of plant species in the monitoring programme for semi-natural vegetation is a cell 40 cm x 40 cm, and this is used in different spatial configurations in the two major elements of the sampling programme. These elements are (i) coarse-grain monitoring which is based on a sample of vegetation selected at random from a series of systematically located grid points and with recording repeated every nine years, and (ii) fine-grain monitoring which is based on a sample of vegetation selected from vegetation types recognisable at the beginning of the programme and with recording repeated at three-year intervals. Coarse-grain monitoring aims to provide a record of broad changes in vegetation at a site, whilst fine-grain monitoring provides the detail which allows changes to be related to the UK's National Vegetation Classification (Rodwell 1991 *et seq.*) The number of fine-grain monitoring locations required is at least two in each vegetation type and this is likely to provide approximately 20 or fewer locations at most ECN sites. The number of coarse-grain monitoring locations is larger, being set at approximately 50, and the detail recorded is less than in fine-grain monitoring.

Provision is also made for additional monitoring of woodlands and of linear features such as hedgerows and at vegetation boundaries, which may shift as a result of environmental change. Some ECN sites are managed in part or in their entirety as permanent grassland or for cereal production, and provision is made for monitoring the annual yield of these crops, using methods which are well established in agricultural research practice.

Method

Vegetation mapping to provide baseline data and sample stratification (VB)

A vegetation map is an essential prerequisite for characterising the vegetation types of each site and thus for selecting areas in which change is to be monitored. An existing vegetation map may be adequate, provided the boundaries between the vegetation types are known to be accurate at the start of the programme. If no such map is available, one should be prepared using recent remotely sensed imagery, land use cover or local botanical survey with ground-truthing of boundaries. It is not necessary at this stage to know the identity of the different vegetation types which can be distinguished. The map should be at a scale of 1:10 000 for sites up to 50 km² in extent, and 1:25 000 for sites over 50 km². Annual patterns of land use at agricultural sites will be recorded on maps and stored in the ECN GIS.

Location

An approximately regular grid, coincident with the National Grid, should be superimposed on the site map, scaled so as to provide approximately 400 sample grid positions. The purpose of the grid is to provide plot locations which are unbiased and re-locatable. If a site includes short-term leys, arable areas, or experimental plots likely to be subject to changing management over the duration of ECN, these should be distinguished on the map under these broad headings, and grid positions falling in these areas excluded from subsequent survey of semi-natural vegetation. Remaining grid positions, falling within semi-natural vegetation, permanent grass and conifer or broadleaf plantations, will be used to characterise the vegetation of the site, together with additional randomly located sample positions placed in other distinguishable vegetation types which are unrepresented or under-represented at the grid positions. There should be at least two sample (infill) points in such additional vegetation types, with no more than 100 in total. A maximum of 500 sample positions will now be available for characterising the vegetation.

If there are existing data of this general kind which can be used to characterise the vegetation types, new sampling can be limited to those areas which were originally under-recorded or where change has obviously taken place since the data were collected.

A 2 m x 2 m plot is centred on each grid and infill point, orientated N, E, S, W using magnetic bearings on the first visit.

V Protocol

88

The centres and corners of plots selected for continued monitoring (see following paras VC, VF, etc) should be permanently marked at this stage. At extensive sites having few landscape features and low-intensity management, it may be essential to use a plot centre marker which can be seen easily from a distance. Plot corners should be marked with buried metal stakes.

Sampling

A species list of all vascular plants, bryophytes and lichens, except those growing on rock or wood, is recorded in the 2 m x 2 m plot, using Tutin *et al.* (1964 *et seq.*), Corley and Hill (1981) and Purvis *et al.* (1993) as standards for nomenclature. Where points fall in peaty pools, streams or lowland ponds, recording will be more difficult, but should nevertheless be attempted unless it results in an unacceptable health or safety risk. Where points fall in woodland or scrub, the trees and shrubs in a 10 m x 10 m plot, centred on the 2 m x 2 m plot and oriented in the same directions, should be listed separately to provide a more representative sample of the canopy and understorey.

Vegetation types identified as a result of this initial survey of the site will be used as a basis for selecting samples for future vegetation monitoring. It is therefore very important that the vegetation types are named by reference to a single classification scheme relevant to all ECN sites. The National Vegetation Classification (NVC) (Rodwell 1991 *et seq.*) provides such comprehensive national coverage of all British semi-natural vegetation, improved grasslands and plantations and is compatible with the European Community CORINE (Co-ordinated Environmental Information in the European Community) Biotopes and Palaearctic Habitat Classifications. Samples can be allocated to NVC vegetation types individually or in groups characterised from the data using a multivariate classificatory technique such as TWINSPAN (Hill 1979). The programs MATCH (Malloch 1991) and TABLEFIT (Hill 1989,1993) provide simple statistical techniques to assist with such matching.

If competence in identifying plants is lacking, then both the original recording and the subsequent monitoring must be subcontracted to skilled surveyors. Sampling must in any case always be timed well within the growing season (June–August at most UK sites) when identification is usually more reliable. Where sampling is in grass managed for silage, recording should precede the first cut.

V Protocol

Coarse-grain vegetation monitoring (every nine years) (VC)

Location

A random selection should be made of 40 of the 2 m x 2 m plots on the regular grid set up for baseline vegetation recording (see VB). Where infill plots exist, up to ten of these plots should be selected randomly in addition, providing a total of up to 50 plots for coarse-grain monitoring. Plots selected for this purpose should be permanently marked (see VB) and are recorded every nine years.

Sampling

Plots should be divided into 25 cells, each 40 cm x 40 cm. In each cell the presence of all vascular plants rooted in the cell is recorded, with the exception of those growing on rock or wood; non-vascular plants are recorded in the same way but in three groups: sphagna, other bryophytes and lichens. This will provide an estimate of local frequency of taxa in each plot. The presence of bare soil, bare rock, litter, dead wood or open water are recorded in the same way.

The altitude, slope, aspect, land use and slope form are noted for each 2 m x 2 m plot (using the terminology of Hodgson 1974), as also are biotic or treatment effects such as grazing and browsing, trampling and dunging by stock or wild herbivores, burning or disturbance, as specified on the recording form which is provided.

If the plot falls within woodland or scrub, then additional monitoring should be carried out as described below (see VW).

Fine-grain vegetation monitoring (every three years) (VF)

Location

Fine-grain monitoring is to be carried out every three years in the TSS and in at least two locations in each of the vegetation types characterised by the methods described above. The locations are chosen to coincide with original grid and infill locations where possible, but otherwise should be selected using randomly selected pairs of co-ordinates. They should not coincide with coarse-grained monitoring locations so as to avoid unnecessary and repeated disturbance to the vegetation. The same plots are recorded on each occasion.

Sampling

V Protocol   A 10 m x 10 m plot is centred around each selected position and oriented N, E, S, W using magnetic bearings, and marked as described above (see VB). Ten 40 cm x 40 cm cells are selected randomly, using a different randomisation for each plot, and marked in their NE and SW corners; each cell is used for recording the presence of all species of vascular plants rooted in the cell, and bryophytes and lichens, except those growing on rock or wood. The presence of bare soil, bare rock, litter, dead wood and open water should be recorded in the same way. Physical and biotic features of the 10 m x 10 m plot are recorded on each occasion, as described above (see VC).

Method Additional coarse-grain monitoring in woodland (every nine years*) (VW)

Location

Where grid and infill samples selected for coarse-grain monitoring fall in scrub or woodland, a 10 m x 10 m plot, centred on the 2 m x 2 m plot, will be used for recording trees and shrubs. The corners of the 10 m x 10 m plot should be marked to aid relocation.

Sampling

Tree and shrub species rooted in the 10 m x 10 m plot will be listed, with a note on whether they are represented (see Appendix II) as canopy dominants (C), subdominants (S), intermediate (I), suppressed (U), shrub layer, saplings or seedlings (Figure 12). A species may be represented in more than one of these categories. Ten cells, each 40 cm x 40 cm, are selected at random using a different randomisation for each 10 m x 10 m plot; the cells are marked, for relocation, in their NE and SW corners. The diameter at breast height (dbh, measured with a tape to the nearest 0.1 cm at a height 1.3 m above the ground) and height (measured to the nearest 0.5 m using a hypsometer or poles) of up to ten trees or shrubs of >5 cm dbh will be recorded*; the individuals chosen are those nearest to the centre points of the ten randomly selected cells. If the plant is multiple-stemmed, the dbh and height of the tallest live stem are measured and the number of stems is counted and recorded. The distance, at a height of 1.3 m, between the approximate centre of the chosen stem and the centre of its associated random cell will be recorded. The measured stems should be marked with paint at 1.3 m, numbered and re-measured on subsequent occasions. If a stem dies between surveys, a replacement stem is selected from the same randomly selected cell using the procedure outlined above.

*Tree diameters are measured every three years

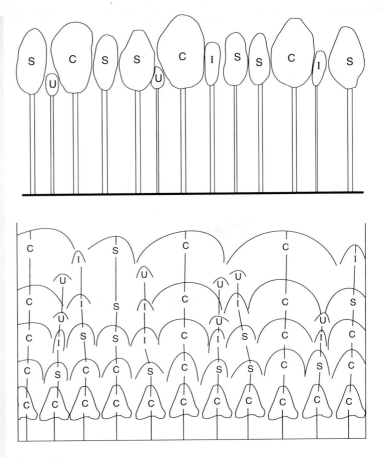

Figure 12.
  i. Relative positions of trees in different crown classes in an even-aged, unthinned pure stand
 ii. Development of different crown classes with advancing age as a result of competition amongst trees originally dominant
       C, dominant; S, subdominant; I, intermediate; U, suppressed

Seedlings will be counted by species in the ten selected 40 cm x 40 cm cells. Seedlings can be individually marked if monitoring of survival is of interest and recording repeated annually if necessary.

Forest health will be assessed annually using UN-ECE (United Nations Economic Commission for Europe) guidelines (Innes 1993).

| **Method** | **Additional monitoring of vegetation boundaries and linear features, including hedgerows (every three years) (VH)** |

Location

Boundaries between the vegetation types characterised and monitored using the methods outlined above can themselves be very sensitive to change. Such boundaries should be identified, from baseline or other surveys, or from aerial photographs, and a number of these should be selected which will adequately represent major discontinuities of the vegetation present at the site. Linear features of particular interest, such as hedgerows and ditches, should also be sampled using the same methodology. Having defined the boundaries to be sampled, one or more transect lines will be located at random and laid out at right angles to each boundary.

Cells of 40 cm x 40 cm are centred on each transect line and are located at suitable regular intervals along the line, spaced as closely and extending as far on each side of the boundary as required; they should be permanently marked at the points where the transect crosses the cell boundaries.

Sampling

The presence of all species of vascular plants rooted in the cell, and of bryophytes and lichens, is recorded, excluding those growing on rock or wood, together with bare soil, bare rock, litter, dead wood and open water, and other physical and biotic factors (see VC). A sketch map should be drawn to show the transect profile and the location of cells along the transect.

Where a boundary shifts during the ECN programme, the transect line and number of cells will be extended accordingly.

Where a hedgerow is to be sampled, species composition of a 10 m length of the hedgerow is recorded in addition. The sample is centred on the transect line described above (see **Location**), ie taking a 5 m length of hedge on both sides of the transect line. Woody species occurring in the whole width of hedge are listed.

| **Method** | **Additional monitoring of permanent grass (four times per year) (VP)** |

Location

At lowland sites the TSS will usually have been selected in an area of permanent grass (>5 years since sowing) which has been managed and utilised in a standard manner for several years, normally by grazing.

V Protocol

93

The TSS should be divided into ten subplots of approximately equal area, within each of which one exclusion cage, covering a minimum area of 3 m^2, is sited at random in mid-March and re-located at random annually.

The existing system of management, including the amount of fertilizer applied and the method of utilisation, should be maintained on the sampled area throughout the ECN programme. All treatments should be recorded, including stocking rates.

Sampling

Species presence will already have been recorded in the TSS in accordance with the protocol for fine-grain monitoring of semi-natural vegetation at three-yearly intervals (see VF).

Grass yield will be recorded in each of the ten exclusion cages. Dry matter yields from the sward will be estimated by mowing under the cages. Four cuts will be taken from the cages using a hand-driven autoscythe, the first in mid-May and again in mid-July, early September and late October. Every effort should be made to make cuts on the same date each year. Although this procedure will not measure the herbage consumed by the stock, it indicates the potential production of the pasture which has been grazed in previous years. The final cut, in October, is intended to coincide with the end of the growing season. Samples will be oven-dried and results expressed as dry matter production/cut and as production ha^{-1} year^{-1}.

Where the permanent grass has to be regularly managed by mowing, every effort should be made to follow the timing described above.

As optional, additional monitoring, it is desirable that one or two sites should look for evidence of change in the potential productivity of grassland by using freshly sown, monospecific swards each year (Corrall & Fenlon 1978).

Method   **Additional monitoring of continuous cereal (annually) (VA)**

Location

Continuous cereal cropping in a uniform area which has been cropped in a standard manner for a significant period is recommended for ECN monitoring. In these circumstances a standard fertilizer treatment will be required; herbicide will be required to deal with grass weed problems in autumn-sown

V Protocol

94

cereal sequences. Spring herbicide treatment should be delayed until after the spring recording of species presence (see **Sampling** below). Fungicides and insecticides will be used as required. All treatments should be recorded.

Species presence will be recorded annually (weeds and crop plants) in at least two 10 m x 10 m plots before the crop elongates and after the spring flush of weeds. The plot locations are chosen to coincide with original grid and infill locations where possible, but otherwise should be selected using randomly selected pairs of co-ordinates.

Estimates of yield will be made within the crop area in close proximity to that used for vegetation monitoring.

Sampling

Apart from its being annual and not triennial, recording of species presence will follow the principles laid down for fine-grain monitoring in semi-natural vegetation (see VF above), ie it will use ten permanent 40 cm x 40 cm quadrats randomly located in each 10 m x 10 m plot.

Twenty plots, each 20–25 m long and the width of the plot combine (minimum plot area 50 m^2), will be harvested. The grain produced on each plot will be weighed, as will a 10 m length of straw swath. Grain yield will be recorded as grain at 15% moisture content; dry matter yield of straw will also be recorded.

As optional extra monitoring, arable crops can be compared in terms of biomass production, for which purpose it will be necessary to aim for sampling at maximum biomass. Winter wheat will therefore be sampled at anthesis by cutting as near as possible to the ground. Twenty plots, each 1 m^2, should be sampled from within a uniformly growing area of at least 50 m x 100 m which is as similar as possible to the area sampled for grain yield.

The dates of standard growth stages of wheat crops at ECN sites should be recorded annually (Tottman & Broad 1987). Assuming that the winter wheat crop is drilled before the end of October, it is recommended that the dates should be recorded on which the following stages are reached:
- GS 31 (start of stem elongation)
- GS 39 (flag leaf emerged)

The dates should be recorded on which 50% of a sample of tillers on a 50 m x 100 m area have reached these stages of development.

Time

Because the size and character of the ECN sites, and the amount of existing remotely sensed, cartographic and survey data are so variable, it is difficult to estimate the time needed to prepare a basic vegetation map.

Estimated times for the collection of initial plot records for characterisation of the vegetation types and for subsequent monitoring are as follows. Estimates are based on an initial array of 500 plots (400 grid + 100 infill), and make some allowance for walking time and bad weather.

Initial recording	30 mins/plot	10 weeks in year 1
Location and marking of monitoring plots	30 mins/plot	2 weeks in year 1
Coarse-grain monitoring	1 hour/plot	4 weeks in years 1,10,19, etc
Fine-grain monitoring	1 hour/plot	4 weeks in years 1,4,7,10, etc
Woodland monitoring	½ day/plot	2 weeks in years 1,10,19, etc

Estimated times required for additional monitoring of hedgerows and linear features, permanent grass and continuous cereal are as follows.

Initial plot establishment	15 days
Hedgerows and linear features	15 days/3 years
Grassland	5 days/year
Continuous cereal	9 days/year

Authors *J.S. Rodwell, J.M. Sykes and M.B. Helps*

References Corley, M.F.V. & Hill, M.O. 1981. *Distribution of bryophytes in the British Isles.* Cardiff: British Bryological Society.

Corrall, A.J. & Fenlon, J.S. 1978. A comparative method for describing the seasonal distribution of production from grasses. *Journal of Agricultural Science*, **91**, 61–67.

Hill, M.O. 1979. *TWINSPAN – a Fortran program for arranging multivariate data in an ordered two-way table by classification of individuals.* Ithaca, New York: Section of Ecology and Systematics, Cornell University.

V Protocol

Hill, M.O. 1989. Computerised matching of releves and association tables, with an application to the British National Vegetation Classification. *Vegetatio,* **83**, 187–194.

Hill, M.O. 1993. *TABLEFIT Version 0.0, for identification of vegetation types.* Huntingdon: Institute of Terrestrial Ecology.

Hodgson, J.M. 1974. *Soil survey field handbook.* Harpenden: Soil Survey.

Innes, J.L. 1993. *Forest health: its assessment and status.* Wallingford: CAB International.

Malloch, A.J.C. 1991. *Running a computer programme to aid the assignment of vegetation data to the communities and sub-communities of the National Vegetation Classification.* Lancaster: University of Lancaster.

Purvis, O.W., Coppins, B.J., Hawksworth, D.L., James, P.W. & Moore, D.M. 1993. *Lichen flora of Great Britain and Ireland.* London: Natural History Museum.

Rodwell, J.S. 1991 *et seq. British plant communities.* Cambridge: Cambridge University Press.

Tottman, D.R. & Broad E.H. 1987. Decimal code for the growth stages of cereals. *Annals of Applied Biology,* **110**, 683–687.

Tutin, T.G., Burges, N.A., Chater, A.O., Edmondson, J.R., Heywood, V.H., Moore, D.M., Valentine, D.H., Walters, S.M. & Webb, D.A. 1964 *et seq. Flora Europaea.* Cambridge: Cambridge University Press.

Appendix I   Siting of vegetation plots and the effects of artefacts

ECN vegetation plots are located according to systematic and then random criteria in order to sample vegetation as objectively as possible. Some of these locations may be in difficult terrain, or may include obstructions or types of land cover which prevent vegetation growth (eg tarmac). The following guidelines should be used in such situations.

- Plots should be laid out so as to follow the ground surface as closely as is practicable. This includes following the profile of ditches, turf banks, bluffs, etc. Descriptive and quality information should be given as indicated in the recording forms.

- If part of the plot is occupied by artefacts such as tarmac

Baseline recording	Record the percentage of the plot occupied by the artefact and therefore unsurveyed. (This is provided for in the Baseline Descriptive/Quality Form)
Coarse-grain recording	If a cell could not be surveyed, use 'Quality Code' 136 (provided for in the Coarse-grain Descriptive/Quality Form for cells)
Fine-grain recording	If five cells or more in the 10 m x 10 m plot are not able to be surveyed, then select randomly another grid plot (if none are available then use random co-ordinates within the vegetation type being sampled). If less than five cells are affected, then re-select those cells using random local co-ordinates

- If part of the plot is obstructed by an 'impenetrable' barrier such as a stone wall

Baseline recording	Survey that part of the plot which falls on the same side of the barrier as the plot centre point. Record the percentage of the plot unsurveyed, as above
Coarse-grain recording	Survey that part of the plot which falls on the same side of the barrier as the

	plot centre point. Record at cell level that a cell could not be surveyed, as above
Fine-grain recording	Select randomly an alternative grid plot so that the whole of the plot falls within the vegetation type being sampled. If no grid plots are available, select a plot using random co-ordinates

- If part of the plot includes a fence, natural boundary, or other linear feature

Baseline and coarse -grain recording	Survey as a normal plot, but record that the feature is present on the appropriate forms
Fine-grain recording	Select randomly an alternative grid plot so that the whole of the plot falls within the vegetation type being sampled. If no grid plots are available, select a plot using random co-ordinates

- Woodland monitoring
Seedlings

If more than half a cell is taken up by a mature tree stem, then select an alternative cell using random local co-ordinates

Appendix II Tree crown classes

1. **Dominant**
   Trees with crowns extending above the general level of the crown cover and receiving full light from above and partly from the side; usually larger than the average trees in the stand, and with crowns well developed but possibly somewhat crowded on the sides

2. **Subdominant**
   Trees with crowns forming the general level of the crown cover and receiving full light from above but comparatively little from the sides; usually with medium-sized crowns more or less crowded on the sides

3. **Intermediate**
   Trees shorter than those in the two preceding classes but with crowns extending into the crown cover formed by dominant and co-dominant trees; receiving a little direct light from above but none from the sides; usually with small crowns considerably crowded on the sides

4. **Suppressed**
   Trees with crowns entirely below the general level of the crown cover, receiving no light either from above or from the sides

5. **Tree**
   A woody perennial plant with a diameter at breast height (dbh; 1.3 m above the ground) >5 cm, typically with a single, well-defined stem carrying a more or less definite crown

6. **Sapling**
   A young tree, no longer a seedling and typically growing vigorously. It has a dbh between 0.5 cm and 5 cm

7. **Seedling**
   A young tree or shrub, grown from seed, from its germination up to the sapling stage, ie with a dbh <0.5 cm

8. **Shrub**
   A woody perennial plant with persistent and woody stem(s). It differs from a tree or sapling, as defined here, in its lower stature and the general absence of a well-defined main stem, ie the side shoots are generally well developed

INVERTEBRATES

Overall aim *To monitor changes in populations of selected groups of invertebrates*

Rationale The invertebrates form a large group in terms of species richness and many of them pose difficult problems for long-term monitoring: sampling for many groups is labour intensive, identification difficult, time-consuming and therefore expensive. However, some groups are ubiquitous, are more readily sampled and identified, and are known to respond to changes in climate, pollution and land use (eg Luff & Woiwod 1995); moreover, in some cases there are already good biological, physiological or ecological data, often from long-term studies, which can provide a background to interpretation. A general policy in considering suitable invertebrate groups for ECN sampling has been to concentrate principally on indicator groups rather than on individual species, and wherever possible on groups where monitoring schemes already exist. · It was desirable to include examples of both herbivores and predators, as well as some populations exhibiting genetic variation, particularly where an environmental link is known. Practical considerations of the availability of expertise and of other resources have also been taken into account so that as wide a range of invertebrates as possible is being sampled within the resources likely to be available within the ECN programme.

Reference Luff, M.L. & Woiwod, I.P. 1995. Insects as indicators of land-use change. In: *Insects in a changing environment,* edited by R. Harrington & N.E. Stork, 399–422. London: Academic Press.

TIPULIDAE

Aim   *To monitor changes in the populations of selected species of Tipulidae larvae*

Rationale   The larvae of craneflies (Tipulidae) are widespread, numerous and easily sampled in soil. They are rarely, if ever, found at depths below 10 cm and sampling does not involve the degree of soil disturbance which would be incurred, for instance, in sampling earthworms. These larvae are mainly plant feeders; they are important food for many other animals and play a key role in terrestrial food webs (Coulson 1962). Because several species might be present, estimates of changes of abundance will be possible on a comparative basis. To ensure that sampling covers the later, larger larval stages, for which both retrieval and identification are easier, it is necessary to sample both in spring and in autumn because, although most species have an annual life cycle, they have different emergence seasons.

Method   Cores are taken from the soil and hand-sorted to remove the larvae. Sampling the larvae from soil cores in April covers the development time of the most common grassland species, *Tipula paludosa*, whilst samples in September will include other species found in woodland and moorland habitats as well as the late spring-emerging grassland species.

Equipment

A metal corer, 10 cm in diameter and with an operating depth of at least 10 cm, will be used.

Location

A sampling site of at least 50 m x 40 m should be selected outside, but preferably close to, the TSS. The sampling is destructive and its long-term effects are mitigated by using as large an area as possible. If both woodland and grassland sites are available, it is desirable to sample each separately as different species will be present. On upland sites it is desirable to select both a relatively well-drained and a wet peat site, as well as a grassland with mineral soil.

Sampling

The sampling area should be gridded into 20 subplots as an aid to taking a stratified random sample which is geographically dispersed across the area. A corer 10 cm in diameter will be used to obtain cores 10 cm deep. Twenty cores should be taken, using pairs of random co-ordinates, one from each of the subplots, in

April and again in September. The cores should be placed in separate, labelled polythene bags and should be hand-sorted within 24 hours. The extracted larvae should be dropped into near-boiling water and then preserved in 70% alcohol, keeping separate those from each core. It should be borne in mind that zero counts may be important for comparison with future counts.

Identification of Tipulidae to species level should be possible by Site Managers, using token specimens provided. Subsequent periodic checking of identifications will also be needed. Adult craneflies collected from the pitfall traps (see IG Protocol) may provide useful confirmation of the presence of some species at a site.

Time

Sampling and sorting	4 days/year
Identification	1 day/year

Author *J.C. Coulson*

Reference **Coulson, J.C.** 1962. The biology of *Tipula subnodicornis* with comparative observations on *Tipula paludosa*. *Journal of Animal Ecology*, **31**, 1–21.

IT Protocol

IM Protocol	# MOTHS

Aim — *To collect and identify daily collections of macrolepidoptera, using a standard light trap*

Rationale

The Lepidoptera (butterflies and moths) is a large order of insects and is among the best known, both taxonomically and biologically. For this and other reasons, such as their phytophagous larval habits, they are a good indicator group for environmental change (Woiwod & Stewart 1990; Woiwod & Thomas 1993). The Protocol involves sampling by light trap of all the macrolepidoptera (large moths) using the methodology of the Rothamsted Insect Survey. This provides a strong background of expertise to ensure that the Protocol requirements are met, as well as an existing extensive national database accumulated over the last 25 years which will form the basis for analysing future trends in populations. The background to the Rothamsted Insect Survey's light trap network is given in Taylor (1986) and in Woiwod and Harrington (1994).

Method

Equipment

A standard Rothamsted light trap (Williams 1948), which requires a continuous power supply, is used.

Location

The light trap should be sited so as to be convenient for daily emptying. Its location should be as near as possible to the TSS but will often be near to laboratories, houses or farms, where help from the occupants can sometimes be enlisted; this is likely to be the only possibility for remote sites, because of the need for a continuous power supply. The trap should be sheltered by vegetation if possible and ideally placed further than 20 m from artificial light sources.

Sampling

Ideally, traps are emptied daily throughout the year but if this is not possible samples should be accumulated, for example at weekends. Currently samples are posted to Rothamsted for identification, unless local expertise is available. All data are lodged in the Rothamsted Insect Survey's existing database as well as in the ECN database. For large sites it is desirable to run more than one trap if suitable operators can be found. Detailed instructions for trap operators are given in Appendix I.

Time

Emptying trap	40 minutes/trap/week, plus travelling
Identification	Up to 15 days/site/year for a very experienced identifier

Author — *I.P. Woiwod*

References

Taylor, L.R. 1986. Synoptic dynamics, migration and the Rothamsted Insect Survey. *Journal of Animal Ecology,* **55**, 1–38.

Williams, C.B. 1948. The Rothamsted light trap. *Proceedings of the Royal Entomological Society of London (A),* **23,** 80–85.

Woiwod, I.P. & Harrington, R. 1994. Flying in the face of change: the Rothamsted Insect Survey. In: *Long-term experiments in agricultural and ecological sciences,* edited by R.A. Leigh & A.E. Johnston, 321–342. Wallingford: CAB International.

Woiwod, I.P. & Stewart, A.J.A. 1990. Butterflies and moths – migration in the agricultural environment. In: *Species dispersal in agricultural habitats,* edited by R.G.H. Bunce & D.C. Howard, 819–202. London: Belhaven.

Woiwod I.P. & Thomas J.A. 1993. The ecology of butterflies and moths at the landscape scale. In: *Landscape ecology in Britain,* edited by R. Haines-Young, 76–92. (Working paper no. 21.) Nottingham: IALE (UK), Department of Geography, University of Nottingham.

The Rothamsted Insect Survey light trap: description and daily operation procedure

The Rothamsted Insect Survey (RIS) light trap (Williams 1948) consists of a stand approximately 1.2 m in height which supports the light source and trap. The overall height is approximately 1.5 m and the trap and stand are approximately 0.6 m in width and depth.

The light source is a 200 watt, clear, tungsten lamp powered by mains electricity. For safety, the supply is protected by a trip switch or Residual Current Device (RCD). The lamp is switched on at dusk and off at dawn each day by an automatic solar dial time-switch which self-adjusts for seasonal variation in daylength.

On entering the trap, insects are killed in a collecting jar which is lined on its inner surface with plaster and into which is absorbed daily approximately 5 ml of 1,1,2,2 tetrachloroethane. The fumes from this chemical render the insect immobile very quickly, thus preventing escape and excessive damage.

The daily routine for sample collection takes approximately five minutes. The recommended procedure, as carried out by staff of the RIS, is outlined below.

Procedure

1. Check that the RCD is in the 'on' position. If it has tripped, ensure that the power supply is on and try to re-set the RCD. If it trips again, replace the fuse in the plug and try again to set the RCD. If there is still a fault, contact an electrician.

2. Check that the clock is set at the correct time (GMT). The solar dial clock adjusts automatically for seasonal variation in daylength. It must not be altered to match changes in local time (eg British Summer Time).

3. By turning the manual switch on the clock, check that the light bulb is working. If it is not, replace the bulb and try again. If it still does not work, contact an electrician. Only 200 watt, clear, tungsten bulbs should be used.

4. Take a collecting jar, which has been dosed with approximately 5 ml of tetrachloroethane and capped with the appropriate jar lid, to the trap and replace the one containing the previous night's sample. To prevent contact with the tetrachloroethane vapour and spillage of the sample, always use the jar lid.

5. In a well-ventilated area or a fume cupboard, tip the sample into one or more (depending on the size of the sample) tissue-lined pill boxes. Ensure that no insects remain in the jar and replace

IM Protocol

the jar lid. Gently fold the tissue over the sample, taking care to avoid crushing the insects, and close the box. Re-dose the jar for use the following day with approximately 5 ml of tetrachloroethane and immediately replace the lid.

6. On the box or boxes containing the sample, write the site name and the date on which the sample was taken. If it was collected on 15 June 1995, the light would have been switched on at dusk on the 14 June and off at dawn on the 15 June. The correct date to be written on the box is therefore 14–15 June 1995. If the sample is accumulated over several nights, the correct date would be that on which the trap was set and that on which the jar was emptied. For example, if the trap ran over the weekend of Saturday 28 and Sunday 29 January, and was emptied on Monday 30 January, the correct date would be 27/30 January 1995. If two boxes are used to accommodate the sample, write the appropriate date and '1/2' on each box. Details of any dates on which the trap was not operating, eg because of lamp failure, should be reported.

Safety   ## Tetrachloroethane

Although alternatives are being sought, tetrachloroethane is the only compound which has the correct properties to enable the rapid killing of insects entering the collecting jar whilst also remaining effective for 24 hours or more. **However, it is toxic by inhalation, skin and eye contact and is assumed to be poisonous if taken by mouth. Prolonged continued high exposure may cause jaundice. The following precautions must therefore be observed.**

1. Avoid breathing the vapour and avoid contact with skin, especially the eyes.

2. When pouring, wear safety spectacles; always work in a well-ventilated area, preferably outside; take the collecting jar to the tetrachloroethane, not *vice versa*.

3. Allow small spillages to evaporate; keep away from the spillage site.

4. Large spillages should be absorbed on to sand and removed outside to evaporate. When doing so, wear safety spectacles, gloves and a respirator.

5. In the case of skin contact, remove contaminated clothing and wash the skin with running water followed by soap and water.

IM Protocol

Electricity

The trap operates by mains electricity. The supply should always be disconnected before a bulb is replaced. No part of the electrical circuit should be exposed without prior consultation with an electrician. The RCD should be tested regularly, according to the manufacturer's recommendations.

BUTTERFLIES

Aim *To monitor the abundance of butterflies using the transect methods of the Butterfly Monitoring Scheme*

Rationale Butterflies are one of the easier insect groups to identify and monitor and are known to respond rapidly to changes in vegetation abundance and quality (Thomas 1991). The Protocol adopted by ECN is that already in use for the national Butterfly Monitoring Scheme, operated jointly by the Institute of Terrestrial Ecology and the Joint Nature Conservation Committee and organised from the Biological Records Centre at ITE Monks Wood (Hall 1981; Pollard, Hall & Bibby 1986). This existing scheme will provide a strong background of information from approximately 15 years of sampling for comparison with ECN data. Analysis of the existing data has already shown interesting changes in the distribution and phenology of individual species (Pollard 1991) and significant relationships between butterfly population size and climate (Pollard 1988).

Method Location

A fixed transect route is set up at each site following the instructions in Hall (1981), and is strictly followed on each sampling occasion. The route is selected so as to be reasonably representative of the ECN site and will often follow existing paths or boundaries and include areas under different management regimes. If necessary, the route should be marked out to ensure that the same route is followed on each occasion. The length of the transect will depend on local conditions but should be capable of being walked at a comfortable, even pace in 30–90 minutes and will therefore usually be 1–2 km. The transect should be divided into a maximum of 15 sections, which may be of different vegetation, structure, or management and which are used as sampling strata. The length of each section is recorded on a map, together with information on habitat types and abundant plants, especially butterfly food plants. Management operations in the vicinity of the transect are also recorded. Changes in these characteristics are noted, so as to assist in interpretation of results.

Method

Recording of the transect takes place weekly between 1 April and 29 September, between 10.45 and 15.45 BST. The temperature should be 13–17°C if sunshine is at least 60%, but if the temperature is above 17°C recording can be carried out in any conditions, providing that it is not raining. At northern, upland sites the appropriate upper temperature is 15°C. The use of more than one recorder would make recording easier at the peak of the season when high numbers of different species occur, or to provide cover for an absent recorder. Transects should be walked by someone with a good knowledge of the British butterfly fauna.

The transect is walked at an even pace and the number of butterflies which are seen flying within or passing through an imaginary box, 5 m wide, 5 m high, and 5 m in front of the observer, are recorded by species for each section of the transect, using the forms provided. Start time is recorded, as are the temperature, percentage sun and wind speed at the completion of the transect. Percentage sunshine is also recorded section by section as the transect progresses.

Time   0.5–1.5 h/transect (depending on transect length) for 26 weeks each year

Author   *I.P. Woiwod*

References   Hall, M.L. 1981. *Butterfly Monitoring Scheme. Instructions for independent recorders.* Cambridge: Institute of Terrestrial Ecology.

Pollard, E. 1988. Temperature, rainfall and butterfly numbers. *Journal of Applied Ecology,* **25**, 819–828.

Pollard, E. 1991. Changes in the flight period of the hedge brown butterfly *Pyronia tithonus* during expansion of its range. *Journal of Animal Ecology,* **60**, 737–748.

Pollard, E., Hall, M.L. & Bibby, T.J. 1986. *Monitoring the abundance of butterflies.* Peterborough: Nature Conservancy Council.

Thomas, J.A. 1991. Rare species conservation: case studies of European butterflies. In:*The scientific management of temperate communities,* edited by I..F Spellerberg, F.B. Goldsmith & M.G. Morris, 149–197. Oxford: Blackwell Scientific.

IB Protocol

SPITTLE BUGS

Aim

*To estimate an index of nymph density of two species of spittle bug (*Neophilaenus lineatus *and*Philaenus spumarius*) and to estimate the proportions of colour morphs of the adults of* P. spumarius

Rationale

Two species of spittle bug, *Philaenus spumarius* and *Neophilaenus lineatus,* are both widespread and common throughout the United Kingdom. Their nymphs, which are xylem feeders, are surrounded by a mass of froth or spittle which the nymphs produce by forcing air into a fluid exuded from the anus. As well as providing some protection from predators, the presence of the spittle makes the nymphs visible and therefore relatively easy to sample. There is a solid background of published ecological work on these species both at lowland and at upland sites (eg Whittaker 1965, 1969, 1970, 1971). In addition to estimating an index of nymph density, the Protocol involves the sampling of colour morphs of *P. spumarius* adults. The colour polymorphism is mainly determined by a series of closely linked genes; it is likely that the proportions of morphs are environmentally determined (Whittaker 1968) and will therefore be good indicators of environmental change.

Method

Populations of *P. spumarius*, which occurs on dicotyledons, and *N. lineatus*, which occurs on monocotyledons, should be monitored annually. Counts of the spittle produced by nymphs of these species are used as an index of nymph density. In addition, the proportions of the different colour morphs of adult *P. spumarius* should be estimated at each sampling location once each year. Monitoring of spittles should preferably be on the TSS but otherwise on an area at the site where they are known to occur.

Equipment

For supplier of sweep nets, please see Appendix II. Quadrats are easily made from wood or metal.

Nymphs

Location

In mid-June 20 quadrats, each 0.25 m², are placed randomly in permanent grassland or in natural vegetation in the vicinity of the TSS. They are marked, their positions recorded, and they are re-visited annually. If a permanent grid is available this can be used in conjunction with random numbers to place the quadrats. Otherwise quadrats can be thrown 'randomly', making sure that no visual reference is made to the vegetation. If monocotyledons and dicotyledons have distinctly different spatial distributions at the site, 20 samples should be taken in each vegetation type, recording *P. spumarius* from the quadrats containing dicotyledons and *N. lineatus* from those containing monocotyledons.

Sampling

The number of spittles occurring in each quadrat is counted separately for dicotyledons (*P. spumarius*) and monocotyledons (*N. lineatus*).

In addition, a random sample of 50 spittles of each species is taken to correct quadrat samples for the mean number of nymphs per spittle. This is done by throwing the 0.25 m quadrat without visual reference to the vegetation or spittle presence and then removing all spittles within the quadrat, keeping separate those spittles collected from monocotyledons (*N. lineatus*) and dicotyledons (*P. spumarius*). This is repeated until enough spittles are found. If spittle density is low, ie most of the quadrats have no spittles, then the first 50 spittles of each species encountered can be used, making sure that there is no bias towards large, conspicuous spittles. These samples are preserved for counting and determination in the laboratory. Nymphs are preserved using a mixture of 70% alcohol, 5% glycerol and 25% water. An absolute minimum sample of 30 spittles of each species is required.

Adults

Location

In late August, sweep net samples of adults are taken in the vicinity of, but not directly on, the TSS. Sampling must be in the same area each year and must be done when the vegetation is dry.

Sampling

A minimum of 50 individual adults of *P. spumarius* is required. Samples should be stored dry, for later determination, in small pill boxes which are then stacked inside a larger container to which naphthalene is added. Each adult is allocated to one of the colour morphs using the methods explained in Appendix II; the numbers of males and females allocated to each morph type are recorded separately and the identified specimens should be stored for future reference and confirmation.

Time	Spittle counts	2 h/site/year
	Sweep netting	1 h/site/year

Author   *J.B. Whittaker*

IS Protocol

References

Whittaker, J.B. 1965. The distribution and population dynamics of *Neophilaenus lineatus* (L.) and *N. exclamationis* (Thun.) (Homoptera: Cercopidae) on Pennine moorland. *Journal of Animal Ecology*, **34**, 277–297.

Whittaker, J.B. 1968. Polymorphism of *Philaenus spumarius* (L.) (Homoptera: Cercopidae) in England. *Journal of Animal Ecology*, **37**, 99–111.

Whittaker, J.B. 1969. Quantitative and habitat studies of froghoppers and leaf-hoppers (Homoptera: Auchenorrhncha) of Wytham Woods, Berkshire. *Entomologist's Monthly Magazine*, **105**, 27–37.

Whittaker, J.B. 1970. Cercopid spittle as a microhabitat. *Oikos*, **21**, 59–64.

Whittaker, J.B. 1971. Population changes in *Neophilaenus lineatus* (L.) (Homoptera: Cercopidae) in different parts of its range. *Journal of Animal Ecology*, **40**, 425–443.

By eye

Divide your group of females into four sections

Pale specimens	Dark specimens with pale fronts
trilineata	Dark specimens with dark fronts

Having sorted into the four groups start with the pale-headed dark group and split it into *marginella, leucocephala* and spotted forms.

Then split the entirely dark group into *lateralis, leucopthalma* and spotted forms.

There will be some specimens in both groups that you will not be able to 'rough sort' by eye.

Using a lens (a x-10 hand lens is quite sufficient)

Check ALL the specimens, group by group, starting with the dark ones which you have just sorted into morphs. You will probably find that one or two have been sorted into the wrong morphs and you can now correct this and also sort the ones which were not separable by eye.

Using the lens, return to the group of pale specimens which you sorted first. Some of these will now turn out to be immature specimens of the darker morphs but under the lens you should be able to see the beginnings of the pattern.

Having dealt with the females, now sort the males in the same way. This will not be quite so easy but having 'got your eye in' with the females will help considerably. Keep the results for females and males separate for entry on the recording form.

In recording your results the most important distinctions to make are into the groups:
    Pale (*typica, populi*)
    Striped (*trilineata, marginella, lateralis* together with *praeusta*)
    Melanics (the rest)
but be as accurate as you can.

Appendix II   Equipment details

Sweep nets (Catalogue no. E679) are obtainable for approximately £50 from:

Watkins & Doncaster
The Naturalists
Four Throws
Hawkhurst
Kent TN18 5ED, UK

GROUND PREDATORS

Aim *To monitor the abundance of ground beetles (Carabidae) and to record features of* Pterostichus madidus *and* Mitopus morio *which may respond to environmental changes*

Rationale In addition to monitoring phytophagous invertebrates such as moths, butterflies and spittle bugs, it has been thought desirable to include a group of predatory species which might amplify changes in their prey or respond in some other, different way to environmental change. The ground beetles (Carabidae) are the obvious choice, being a group which is taxonomically tractable and for which an appreciable body of reliable biological information already exists (eg Thiele 1977). Pitfall trapping has been used extensively and successfully for this group, and a well-developed protocol already exists (M.L. Luff, pers. comm.). Even though it is known that problems exist in interpreting data from such traps, active programmes of standardisation are in progress. Analyses of existing data have already shown that carabids can be sensitive indicators of changes in management (eg Eyre *et al.* 1989). Many beetle species are known to be sensitive to temperature changes (Thiele 1977) and one species, *Pterostichus madidus*, has leg-colour morphs which may be sensitive to climate change (Terrell-Nield 1992); morphs of this species will be recorded separately. The ubiquitous harvestman, *Mitopus morio*, which has features which may respond to environmental change, will also be collected from the pitfall traps.

Method The method will, in general, conform with that developed by Dr M.L. Luff of the University of Newcastle upon Tyne. At each site, in or adjacent to the TSS, a pitfall trapping system should be instituted to sample Carabidae (ground beetles) and the ubiquitous opilionid (Harvestmen) *Mitopus morio*. Adult crane-flies caught in the traps should also be retained in order to check their identification against the larvae extracted in soil cores (see separate Protocol for Tipulidae). It will be necessary to measure the length of the second femur of *M. morio*; this changes with altitude and latitude and may react to environmental change. **All** carabid beetles (not just *Carabus* spp.) are of interest.

Equipment

The pitfall traps are polypropylene cups, 7.5 cm diameter x 10 cm deep (see Appendix I for supplier). A wire netting cage, made from chicken wire with a mesh size approximately 15 mm x 20 mm and approximately 70 mm high, is clipped to the rim of each trap so as to reduce the number of small mammals inadvertently caught in the traps. Each trap should have a cover which can be made cheaply from 5" diameter plant pot saucers and galvanised wire.

This helps to prevent heavy rain from flooding the traps, keeps birds from interfering, and helps in their re-location.

Location

Three transects should be selected, preferably within different vegetation types and including the TSS. Transects should not be placed where cattle have access. Ten pitfall traps should be established in each transect, with 10 m spacing between the traps. If possible, transects should be approached from different directions when attending to traps so as to avoid trampling damage to the vegetation. If trampling is having an obviously deleterious long-term effect on the vegetation, a trapline can be moved each year but returned to the original position every three years. Any movement of traplines should be recorded and information sent to the Central Co-ordination Unit.

Sampling

The traps are set out on the first Wednesday in May and are then emptied and replaced fortnightly for 13 sampling periods, until the end of October. Blue antifreeze is used as preservative; this can be bought from wholesalers in 25 litre drums at a cost of about £30; it is decanted into 1 or 2 litre bottles for use. The undiluted preservative is poured into each trap to a depth of about 3 cm; if diluted by rain, the catch remains in the heavier, undiluted preservative at the bottom of the trap.

Pre-filled replacement traps with marked lids are taken to the sample site. Each trap to be replaced is removed from the hole and the new trap with its lid is placed in the hole, its seating is checked, and its lid removed and placed on the trap containing the catch. Traps may be stored at 4°C prior to emptying.

The catch should be stored at 4°C prior to sorting, but if sorting cannot be undertaken in less than three days after collection the catch should be transferred to 75% alcohol to avoid deterioration of the sample material. To sort, the contents of the trap are poured through a 1 mm mesh, 10 cm diameter test sieve (see Appendix I for supplier). The antifreeze is collected in a beaker placed beneath the sieve and is then poured back into a 5 litre aspirator for future use. The traps are rinsed with water to ensure that any adhering dirt which may contain beetles is washed through the sieve. The sieve is then inverted over a 10 cm diameter crystallising dish and the contents are washed into the dish with water. The taxa of interest are extracted carefully on a large white plastic tray under good light and adequate ventilation into individually labelled vials containing a mixture of 70% alcohol, 5% glycerol and 25% water. Particular care must be taken not to damage or remove legs of *Mitopus morio* specimens which are needed for measurement. The catch is labelled by writing the ECN site identification Code, trap number and collection date on a piece of paper which is then placed in an empty, plastic-stoppered 5 cm x 1.2 cm specimen tube which is dropped into the storage bottle. The catch can now be stored for subsequent identification. Records of the catch from each trap should be kept separate until after identification, when the catch from ten traps can be bulked and stored.

Time

Emptying traps	1 h/month
Identification (for an experienced identifier)	10 h/month

Authors

I.P. Woiwod and J.C. Coulson

References

Eyre, M.D., Luff, M.L. Rushton, S.P. & Topping, C.J. 1989. Ground beetles and weevils (Carabidae and Circulionidae) as indicators of grassland management practice. *Journal of Applied Entomology,* **107**, 508–517.

Terrell-Nield, C.E. 1992. Distribution of leg-colour morphs of *Pterostichus madidus* (F.) in relation to climate. In:*The role of ground beetles in ecological and environmental studies,* edited by N. Stork, 39–51. Andover, Hampshire: Intercept Ltd.

Thiele, H-U. 1977. *Carabid beetles in their environments.* Berlin: Springer-Verlag.

Appendix I   Equipment details

Suppliers   Pitfall traps can be obtained from:

A W Gregory & Co Ltd
Glynde House
Glynde Street   *Tel: 0181 690 3437*
London SE4 1RY, UK   *Fax: 0181 690 0155*

Test sieves are obtainable from:

Endecotts Ltd
9 Lombard Road
London SW19 3TZ, UK

VERTEBRATES

Overall aim   *To monitor changes in populations of selected groups of vertebrates*

Rationale   It is recognised that different organisms are likely to 'filter' the environment in different ways and that, to provide relevant data, an environmental monitoring programme should, ideally, focus on a range of organisms, eg endotherms as well as ectotherms, and organisms with different generation times (Whittaker 1990).

There are many wild, breeding vertebrates in the UK which could be included in this range of organisms and whose populations are likely to be affected by changes in climate, land use and pollution. However, there are considerable difficulties associated with attempting to assess their numbers and this accounts for the general dearth of information on population trends in most vertebrates. Birds are an exception in that they are relatively easy to observe, most are diurnal, and they are identifiable both by sight and by sound. Mammals are more problematical, often being most active at dawn, dusk, and during the night, when observation is most difficult, but an index of their local populations can in some cases be obtained by counting droppings, eg in rabbits and deer, both of which are widespread herbivores. Bats, though crepuscular, can be counted by electronic interception and recognition of their echo-sounding. Frogs are ubiquitous predatory amphibians and, although adult populations are difficult to monitor, it is possible to assess significant changes in populations by counting egg masses.

Reference   Whittaker, J.B. 1990. Technical aspects of detecting change: animals. In: *Environmental Change Network (ECN). Report of a Workshop on the Terrestrial Network,* edited by O.W. Heal, J.M. Sykes & G. Howson, 32–36. Grange-over-Sands: Institute of Terrestrial Ecology.

BI Protocol	**BIRDS**

Aim   *To record the annual distribution and abundance of breeding birds within selected areas of ECN sites*

Rationale   Observations of the devastating effects of agricultural pesticides on bird populations during the late 1950s and 1960s strengthened the view that only objective, regular and scientific monitoring of birds can give an insight into the changing situation of different species. It was also realised that such monitoring can also reflect perturbations affecting the wider environment which ultimately threaten man. Birds are relatively easy to observe and are thus good subjects for a monitoring programme. Moreover, the large number of bird species, over 200, which breed in the United Kingdom have different feeding patterns and occupy many different habitats, making it likely that at least some species will react to particular environmental changes of whatever type. It is well recognised that bird populations are affected by many man-induced factors in addition to pesticides and other agro-chemicals. Land use changes such as drainage and afforestation affect both nesting sites and food availability. Birds are also affected by variations in climate and in particular by periods of severe weather, either in the UK, for species over-wintering here, or in other countries for migratory species.

Whilst it is unrealistic to expect monitoring schemes to be able to identify the causes of population changes, they can be expected to provide an indication of the factors most likely to be responsible and allow unlikely hypotheses to be rejected quickly (Baillie 1990).

Method   The British Trust for Ornithology (BTO) is responsible for organising surveys and providing annual indices of population changes for British breeding bird species. Until recently there have been two major schemes, the Common Birds Census (CBC) and the Waterways Bird Survey (WBS), both of which use a territory mapping method and involve fieldworkers in making about ten visits to their usually subjectively chosen survey plot throughout the breeding season. The CBC (Marchant 1983) has been used for farmland plots, usually areas of at least 40 ha, and semi-natural woodland areas of at least 25 ha, since 1962 and is appropriate for many ECN sites, data from which can be placed in the wider regional and national context. Some ECN sites already had CBC surveys in operation when the ECN programme was initiated in 1993.

The CBC method, described below in more detail (see BC Protocol), is inappropriate for some bird species, particularly the waders, which breed on moorland. Some ECN sites have large areas of moorland and at these it has been necessary to adopt an alternative survey method devised for monitoring waders in the British uplands, supplemented by CBC methods for passerine species. This method is described below (see BM Protocol).

As a result of field trials carried out over the last several years, the BTO is in the process of replacing the CBC with a new census method based on counting breeding birds observed in randomly selected 1 km squares of the National Grid (BTO 1995). This new method, the Breeding Bird Survey (BBS), is therefore more representative of the countryside as a whole and is less labour-intensive than CBC, both in the field and at the analytical stage. The BTO intends to run the CBC and BBS in parallel for several years so as to maintain continuity and to provide a smooth transition from one system to the other.

ECN wishes to retain the advantages of linking to the national scheme for censusing breeding birds and therefore plans to convert to the BBS scheme, which will be appropriate for all ECN sites. In 1996, ECN will start a five-year period of overlap, using both CBC and BBS methods at the ECN lowland sites, and both the ECN moorland breeding bird method (see BM Protocol below) and BBS at upland sites. The BBS method, described below (see BB Protocol), is thus expected to completely supersede existing methods at ECN sites in the year 2001.

All observers using any of the methods described below must be competent to identify readily by sight and by sound all species likely to occur at an ECN site.

Author   *J.M.Sykes*

References   Baillie, S.R. 1990. Integrated population monitoring of breeding birds in Britain and Ireland. *Ibis*, **132**, 151–166.

British Trust for Ornithology. 1995. *Breeding Bird Survey instructions.* Thetford: BTO.

Marchant, J. 1983. *BTO Common Birds Census instructions.* Tring: British Trust for Ornithology.

BI Protocol

Common breeding birds

Aim

To record the annual distribution and abundance of breeding birds within selected areas of ECN sites

Method

The Common Birds Census, operated by BTO, was started in 1962 following trials in 1961, with the aim of monitoring bird population numbers, chiefly on farmland where the growing use of agricultural chemicals and the accelerating rate of hedgerow destruction was causing particular concern. Later other habitats, notably woodland, were included in the scheme. CBC uses a mapping method in which a series of visits are made to all parts of a defined plot during the breeding season, and contacts with birds by sight or sound are recorded on large-scale maps. Information from the series of visits is combined to estimate the number of territories found. Maps of the same plot from different years can be used to assess changes in species densities and to relate these to changes in habitats. CBC results also provide indices of population change for those species which are sufficiently numerous to provide large samples when considered across the sites participating in the scheme. At ECN sites the method will provide information on year-to-year changes in species abundance and allow the sites to be linked to equivalent regional and national patterns.

The method described below summarises the procedures set out in the instructions to BTO observers (Marchant 1983), from which publication further details may be obtained.

Location

The census is carried out in a plot which, for CBC purposes, should be a minimum of 40 ha in farmland (arable, horticultural or grazing land except unenclosed sheepwalk) and 10 ha in woodland (semi-natural broadleaved and mixed woodland, excluding parkland, scrubby heathland and even-aged coniferous plantations). These areas should also be targets for ECN sites, though it is recognised that it will sometimes be impossible to achieve the minimum size. Edge effects, which give rise to inflated estimates of territory density, should be minimised by reducing the edge/area ratio as far as practicable.

Plot boundaries must be clearly discernible features, such as permanent landscape features or an artificially marked grid.

The boundaries of the plot are marked on a 1:2500 map, together with internal detail such as tracks, buildings, hedges, isolated trees, and other easily distinguishable features. If there are insufficient natural features to enable the observer's position to be judged accurately, the area should be gridded at 50 m intervals and the grid positions marked so as to be easily visible and semi-permanent.

Sampling

Habitat recording

Habitat recording enables bird populations to be related to habitat features and to changes in those features. Habitat features of the plot, and extending for 50–100 m beyond the plot boundaries, are recorded on a map before the start of the first year's recording. Important changes in habitat which occur either during the recording year or from year to year are mapped separately. Instructions for habitat description are provided (see page 174).

Frequency and timing

Ten visits should be made between mid-March and late June at southern sites, though these times will need to be adjusted further north. Visits should be spaced evenly through the season and weekly visits are ideal. A complete visit will take approximately 3–4 hours. It is suggested that eight morning visits, starting before 0900 BST, should be combined with two early season evening visits, starting after 1700. Cold, windy and wet days are to be avoided and on particularly fine days an early start is advisable.

Bird recording

The objective of the observer is to mark on the map the location and movements of every bird present or flying over during a visit, but to record each individual only once. A standard list of conventions for species and activities is provided (see pages 177–178) to allow clear and unambiguous recording by observers. Some species are treated differently, as follows.

1. Grey heron (*Ardea cinerea*), rook (*Corvus frugilegus*), sand martin (*Riparia riparia*), feral pigeon (*Columba livia*), and all gulls and terns: if nesting, active nests are counted or estimated and recorded on the map; if present but not nesting, their presence is noted on the margin of the relevant visit map.

2. Wood pigeon (*Columba palumbus*), swift (*Apus apus*), swallow (*Hirundo rustica*), house martin (*Delichon urbica*), magpie (*Pica pica*), jackdaw (*Corvus monedula*), carrion crow (*Corvus corone corone*), house sparrow (*Passer domesticus*) and starling (*Sturnus vulgaris*): these species are censused by a nest count on the plot, or by a combination of nest counts and normal registrations.

3. Fieldfare (*Turdus pilaris*), redwing (*Turdus iliacus*), brambling (*Fringilla montifringilla*) and other common winter visitors will be ignored.

Coverage of the plot should be as even as possible, but more time should be allowed for areas where bird density is higher. The direction and starting point of the route should be varied between visits. A code identifying the visit, the date, starting time, and weather are recorded at the start of each visit; finishing time and the extent of plot coverage are recorded subsequently.

In farmland, progress can be quite fast because the number of birds detectable from any one point is usually rather limited, but the route should take the observer at least once along every major internal hedgerow, as well as completely around the perimeter of the plot. In woodland, a route should be followed which takes the observer within 50 m of every part of the plot at least once during each visit, and a compass may be necessary to enable the observer to follow a marked grid line. The majority of contacts in woodland will be by sound.

Separate species maps are compiled at the end of the season, combining the information from each visit to show the location of birds of the same species on different visits. These are subsequently analysed to show the number of territories occupied by different species.

Author   *J.M. Sykes* (from Marchant 1983)

Reference   Marchant, J. 1983. *BTO Common Birds Census instructions.* Tring: British Trust for Ornithology.

Moorland breeding birds

Aim *To record the annual distribution and abundance of breeding birds within selected areas of ECN sites*

Method Research on moorland bird populations indicates that the number of breeding pairs can be assessed with confidence for the waders, but that, among the passerines, twite (*Acanthis flavirostris*), skylark (*Alauda arvensis*) and meadow pipit (*Anthus pratensis*) present some difficulties. Most of the changes in breeding numbers of moorland birds seem to stem from differences in habitat conditions, weather conditions and food availability, though much debate persists about the additional or compounding effects of predation, parasitism and disease. The methodology described below provides a well-tested means of assessing numbers of breeding birds on moorland.

Licences to disturb breeding birds listed under Schedule 1 of the Wildlife and Countryside Act (1981) should be obtained from English Nature, Scottish Natural Heritage, Countryside Council for Wales or the Department of Agriculture for Northern Ireland.

Location

The survey is based on a representative group of 1 km^2 units of the National Grid. The area to be surveyed will depend on the size of the ECN site and on the consistent availability of staff to carry out this activity on the same area each year. A survey area of 4 km^2 (2 km x 2 km) may be expected to require at least one man-day's work per visit. On enclosed land this is approximately equivalent to a search intensity of one minute per hectare. Each of the 1 km^2 units to be monitored is divided into four sub-squares, each 0.5 km x 0.5 km, which are marked so as to be re-locatable in future surveys. This eases navigation and ensures that all parts of the kilometre square(s) receive equivalent amounts of search effort.

Sampling

Frequency and timing

Each subsquare is surveyed twice during spring/summer by a single observer. The first visit should take place between early-April and mid-May. The second visit should take place between mid-May and end-June. These times may need to be changed, being earlier if there is an early spring and later if spring is late. The survey should be carried out only between 0830 and 1800 BST, thus avoiding periods when the detection of birds is unreliable. The survey should be carried out when wind strength is less than Beaufort Scale Force 5; there should be good visibility, and no persistent precipitation.

Mapping method

Within each subsquare 20–25 minutes should be spent surveying the entire subsquare thoroughly by walking about it in such a way that all parts of the square are approached to within 100 m. On the second visit, subsquares should be visited in the reverse order to that used on the first visit. The observer should stop at regular intervals, scan and listen. Attention is focused on obtaining proof of breeding (see below) and on distinguishing individual pairs of birds.

Apart from red grouse (*Lagopus lagopus scoticus*), skylark, meadow pipit, twite and carrion crow, the locations and activities of all species should be recorded separately (see Chapter 3, page 173) for each of the two visits on a 1:25 000 or 1:10 000 map using the notations of Marchant (1983); the date and starting time for each subsquare should be recorded on the back of the map. Birds are considered to be breeding if any of the following activities are observed:
- song or display;
- birds carrying nesting material;
- nests or young birds found or chicks heard;
- adults repetitively alarmed, indicating nearby nests or young;
- adults giving distraction displays;
- birds carrying food;
- birds in territorial dispute.

All data are transferred from field survey maps to summary visit maps. Where several individuals are present in an area, and it is impossible to determine the number of pairs they represent in the field, individuals are judged to be representative of different pairs only if the distance between them is greater than 500 m (or greater than 200 m for dunlin (*Calidris alpina*) and passerines). In these instances, where two individuals are considered to constitute a 'pair' of birds, the pair's location is placed centrally between the two individuals.

Individual red grouse, skylarks, meadow pipits, twite and carrion crows are counted within each 1 km square only on the first visit.

Population estimates

Maps produced on both visits are considered together in assessing the records for population estimates and in producing distribution maps. For all species other than dunlin and passerines, breeding pairs are considered to be separate from one another only if more than 1 km apart on different visit maps. For dunlin, this distance is reduced to 500 m, and for passerines it is

BM Protocol

129

reduced to 200 m. Where pairs are judged to be the same (ie less than these distances apart), their locations are mapped as being half-way between the mapped observations of pairs on the two individual visit maps.

The distribution maps can be related to habitat and topographical information collected for the site.

A full description of the method is provided by Brown and Shepherd (1993).

Authors *D.B.A. Thompson and A.F. Brown*

References **Brown, A.F.** 1991. *An annotated bibliography of moorland breeding bird and breeding wader surveys, 1970–1990.* (Report no. 8.) Peterborough: Joint Nature Conservation Committee.

Brown, A.F. & Shepherd, K.B. 1993. A method for censusing upland breeding waders. *Bird Study,* **40,** 189–195.

Brown, A.F. & Stillman, R.A. 1993. Bird habitat associations in the eastern Highlands of Scotland. *Journal of Applied Ecology,* 30, 31–42.

Marchant, J. 1983. *BTO Common Birds Census instructions.* Tring: British Trust for Ornithology.

Thompson, P.S. & Thompson, D.B.A. 1991. Greenshanks (*Tringa nebularia*) and long-term studies of breeding waders. *Ibis,* **133** (Suppl.1), 99–122.

Breeding birds

Aim
To record the annual distribution and abundance of breeding birds within selected areas of ECN sites

Method
The Breeding Bird Survey is a key component of the Integrated Population Monitoring Programme operated by BTO. It aims, *inter alia*, to provide information on year-to-year and longer-term changes in population levels for a wide range of breeding birds across a variety of habitats throughout the United Kingdom. The scheme was designed to overcome the limitations of its forerunners (CBC and WBS) by:

i.   selecting sample sites randomly,
ii.   increasing the number of sample sites,
iii. using counts instead of mapping territories, thus
iv.   reducing the sampling effort at each site, whilst
v.   improving representativeness.

The method described below summarises the procedures set out in the instructions to BBS observers (BTO 1995), from which publication further details may be obtained.

Location

At sites where the CBC method (BC Protocol) has been used previously or is still in use during the overlap period, at least one 1 km square, but preferably more, should be selected at random from those squares containing part of the former survey area. At moorland sites, where the BM Protocol has been used previously or is still in use during the overlap period, the same squares should be used as were used previously. At sites where neither BC Protocol nor BM Protocol has been used previously, at least one 1 km square, but preferably more, should be selected at random from all available squares.

Sampling

Transect establishment

Birds are counted on transect lines. In each 1 km square there should be, ideally, two parallel transect lines, oriented north/south or east/west, each 1 km long. Transect lines should be 500 m apart and 250 m from the edge of the square. Each transect line is divided into five equal sections of 200 m, to provide a total of ten consecutively numbered sections. The start of each section should be marked, using either permanent landmarks or easily visible temporary markers. In practice, the transect line may need to deviate from the ideal because of the need to avoid barriers and this is acceptable if the diversion does not result in the two lines being closer together than 200 m. Minor intrusions into adjacent squares are also acceptable where they provide the only practical way to carry out the survey. The exact route taken by the transect lines is marked on a map. The same route is followed every year.

Habitat recording

Habitat recording enables bird populations to be related to habitat features and to changes in those features. Habitat features are recorded separately at the beginning of each recording year for each 200 m section of the transects, taking into account habitats occurring within 25 m of each side of the line. A form is available for the recording of habitat features, identified by code numbers (see page 180). Important changes in habitat which occur during the recording year are also noted on the recording form.

Habitat codes allow the description, for each section, of both the predominant habitat, termed the first habitat, and a second habitat if this exists. For each habitat one habitat code is chosen from each of levels 1 and 2, and up to two from levels 3 and 4 of the habitat classification.

Where the actual transect route differs from the ideal of two parallel transects, the average distance from the ideal is estimated and an additional, separate, list of habitat codes for the ideal transect is also recorded.

Frequency and timing

In the lowlands of southern Britain, the main part of the breeding season, roughly between 1 April and 30 June, should be divided into two counting periods, early-April to mid-May and mid-May to late June, and one visit should be made in each period. Visits should be at least four weeks apart. The first should coincide with the main activity period of the resident breeding birds in the area, whilst the second should take place after the arrival of the latest migrant breeding birds. Visits should be made later at sites which are further north or at higher altitudes.

Counts should be made in the morning, starting ideally between 0600 and 0700 BST, and no later than 0900. The starting time should be consistent both within a breeding season and between years.

Counting should not be attempted in conditions of heavy rain, poor visibility or strong wind, and prevailing weather conditions should be recorded on the forms provided.

Bird recording

All birds encountered whilst the observer walks along the two transects are recorded. The first transect line (sections 1–5) is walked at a slow and methodical pace, starting from the chosen starting place, and noting the starting time of the transect. After completing the first transect line, bird recording stops, the time is recorded and the observer moves to the start of the second transect line (sections 6–10), where the time is recorded and bird recording re-starts. An average visit should take approximately 90 minutes.

The observer should pause briefly and listen for bird songs and scan for birds flying overhead whilst walking along the two linear transects. All birds seen or heard are recorded on the field recording forms in the appropriate transect sections, 1–10, and in one of the following four categories:

1   within 25 m either side of the line;
2   between 25 m and 100 m either side of the line;
3   more than 100 m either side of the line, including birds outside the 1 km square boundary; or
F   birds in flight only (at any distance).

The distances are estimated at right-angles to the line and observers should familiarise themselves with estimating the 25 m and 100 m distances before beginning fieldwork. A bird seen 200 m ahead of the observer but close to the transect line should be recorded in Category 1. Category F relates to birds flying over. If a bird is seen to take off or land, it should be recorded in the appropriate distance category (1–3) at that position. The same individual bird should not be recorded twice.

Standard BTO species codes should be used (see page 177) and this requires the observer to be familiar with the codes of the species most likely to be encountered. Juvenile birds should be distinguished from adults in species where this is possible.

Birds nesting in dense colonies within the square (rook, sand martin and gulls) will not be adequately censused by the standard method and observers should count or estimate the number of nests in the whole 1 km square. Colony counts should be conducted separately from the transects and only for the species listed above.

Following each field visit, observers should produce summaries of the number of adult birds of each species seen in each distance band of the ten sections of each transect (see page 181).

Author   *J.M. Sykes* (from BTO 1995)

Reference   **British Trust for Ornithology.** 1995. *Breeding Bird Survey instructions.* Thetford: BTO.

BI Protocol

BATS

Aim *To assess bat activity and usage of habitat types, and to monitor changes in bat numbers at ECN sites*

Rationale There are 15 species of bat, the only flying mammal, which are considered native to Britain where they are ubiquitous predators of insects. Bats are known to be sensitive to changes in the environment, their numbers having declined significantly during the present century through a combination of loss of habitats, roost sites and food supply, and by increased frequency of disturbance. Bats are therefore believed to be useful indicators of environmental change. Considerable national and international interest in and concern for the future of bat populations has led to a number of survey and monitoring schemes, including a recent survey in Britain (Walsh, Harris & Hutson 1995). Current methodology is somewhat limited in the amount of information which it can provide on the precise relationships between population levels and environmental change; nevertheless, by linking ECN results to those from more widespread monitoring programmes these limitations can be minimised.

Method The method is based on that used in a Bats and Habitats Survey organised by Prof S Harris and colleagues at the University of Bristol for the Joint Nature Conservation Committee. It uses a tunable bat detector to record bats in 1 km squares, noting the positions along a pre-determined transect of all bats seen or heard. Habitat features along the transect, and changes in those features, are also recorded using a standard list.

Location

One or more kilometre squares are selected at the site. The selection need not be at random, and a square which is reasonably typical of the ECN site should be used. Selection should be based on the practicality of safely walking along a transect which crosses the square in darkness.

Using a map at 1:25 000 or 1:10 000, the square is divided into two halves with either a north/south or an east/west line, depending on which direction is easier to walk. For each half of the square, the transect should go from the middle of one edge to the middle of the opposite edge by a reasonably direct but practical route. Where linear features such as field boundaries, hedges, streams or fences can be used these should be followed, but where these do not exist a straight line transect should be used. The chosen route should be marked on the map, the starting and ending positions of the two halves (A–B and C–D) being labelled, and the observer should become familiar with the route during daylight as it will not be as obvious in darkness. In areas where hazards such as ditches, cattle, etc, may pose a problem in the dark, it may be desirable for the observer to have a companion. In any case suitable safety precautions should be taken. In planning the transect, the time taken will need to be checked, bearing in mind that both halves of the transect must be walked on the same night, and that the whole transect should be completed within 45-90 minutes. It should also be remembered that the walk will take longer in darkness and that extra time will be needed when stopping to record any bats seen or heard.

Sampling

Habitat recording

The transect is likely to follow linear features and boundaries between habitat types. Although the observer will not walk through the middle of crops, it will be possible to walk straight across a pasture, rough grazing, moorland, etc, and in such circumstances there is no need to follow any linear features. Having selected the transect, any linear habitat features from the list provided (see page 183) are marked on the map, as are other habitats occurring within about 20 m on either side of the transect/ linear feature being followed. If walking through a uniform habitat, simply record the same habitat feature on both sides of the transect. The habitat features are recorded on a separate 1:10 000 map using coloured crayons, and identified with the habitat code numbers given on page 183. They should be updated annually.

Frequency and timing

The transect should be walked four times in each year, once in each of the following three-week periods:

> 15 June – 6 July
> 7 July – 27 July
> 28 July – 17 August
> 18 August – 7 September

Starting 30–45 minutes after sunset, the transect should be walked steadily at the observer's chosen pace and extra time should not be spent in areas which might be thought especially good for bats. The starting point of the walk should be rotated, if this is practicable, so that it is different on successive occasions. The observer should aim to complete the whole of the transect in 45–90 minutes, including the time taken to get from one half of the transect to the other, during which bats are not recorded

Surveys should not be carried out when rain is heavier than a light drizzle or when there are strong winds, because bats may not be flying in these conditions.

Bat recording

The Batbox III tunable bat detector is recommended for use during the survey; it is capable of recording significantly higher activity than other available models. If the observer is inexperienced, the output from the detector should be connected to a tape recorder so that identification can be improved and confirmed by comparison with a tape of bat calls which is available commercially. The detector should be tuned to 45 kHz and left on this frequency, as this will pick up the largest range of bats. The frequency should not be changed during the walk.

A separate copy of the map should be used for each evening walk, even if no bats have been recorded on previous occasions. At the start and end of the survey, air temperature should be measured, together with cloud cover on an eighths scale (clear sky= 0/8, total cloud cover=8/8). The time at which the walk is started and ended, and the start and end position of each transect are recorded. When a bat is seen or heard, the appropriate symbol (see Chapter 3, page 182) is recorded on the map and the walk continues immediately, with no pause to listen for bats. A circle is drawn on the map where each bat is detected. If the species can be identified with certainty to one of the groups listed, the appropriate letter code is added to the left. If a feeding buzz is heard, add a letter 'f' on the right of the circle, and if it is heard echo-locating but a feeding buzz was not recorded, add a letter 'h'; there may be instances in which both suffixes are attached. Bats which are seen, but not heard on the bat detector, should be recorded with a circle and a letter 's' to the right.

These categories are not mutually exclusive and more than one may be recorded, eg:

Of	bat of unknown species – feeding buzz heard
POs	long-eared bat seen, not heard
Oh	bat of unknown species heard echo-locating
NnOsh	noctule seen and heard echo-locating

Each bat pass consists of an unbroken stream of ultrasonic calls, or for as long as a bat is continuously in sight. When the symbol is marked on the map, the observer should be sure that it is clear in exactly which habitat the bat was recorded. If there is a possibility of confusion, a note should be added at the side of the map for clarification. If any bat roosts are known or suspected in the area, these should be recorded on the map.

Authors   *A.L. Walsh, S. Harris and A.H. Hutson*

BA Protocol

Reference   Walsh, A.L., Harris, S. & Hutson A.H. 1995. Abundance and habitat selection of foraging vespertilionid bats in Britain: a landscape-scale approach. In: Ecology, evolution and behaviour of bats. *Symposia of the Zoological Society of London,* **67**, 325–344.

Appendix I   Equipment required

Equipment   Batbox III tunable bat detector

Supplier   Stag Electronics (manufacturers)
1 Rosemundy
St Agnes
Cornwall TR5 0UF, UK   *Tel: 01872 553441*

RABBITS AND DEER

Aim   *To monitor broad changes in the populations of rabbits and deer*

Rationale   Rabbits (*Oryctolagus cuniculus*) and deer (red (*Cervus elaphus*) and roe (*Capreolus capreolus*)) are the most common wild herbivores in the United Kingdom where they have a considerable effect on vegetation structure and on plant diversity over very large areas of both grassland and woodland. There are several examples of ecosystems in which these animals are the principal determinants of plant composition and structure, and therefore of habitat for other organisms. Whilst it is important to know whether numbers of either or both of these herbivores have changed appreciably, there are no practicable methods of making direct measures of their population size, and an index method based on dropping counts is therefore used to estimate relative abundance. The proposed recording method is expected to detect changes similar to those which have occurred in the past, such as the dramatic fall and slow recovery of rabbit numbers following the introduction of myxomatosis.

Method   **Location**

It is recommended that the same transect used for ECN butterfly monitoring (IB) should be used, extending it where necessary to a length of 2 km. A second transect should be established which covers major habitat types present at the site but which are not encountered on the butterfly transect. This transect should be of a length appropriate to the conditions; at agricultural sites it should be sited so as to include a field centre.

Sampling

In late March, and again in late September, droppings are counted along the pre-determined transects to record the relative abundance of rabbits (and deer if common); it should be borne in mind that zero counts may be important for comparison with future counts. The existing butterfly transect will already have been divided into not more than 15 sections based on discrete habitat types or on subdivisions of the same habitat which are being managed differently, and any extension should be divided in the same way. The length of each section of the transect should be recorded and the habitat type in each section should also be described. The second transect should be subdivided and recorded similarly. All droppings within 1 m of the transects on both sides are cleared two weeks before recording is to take place. At the time of recording the transects are walked again and droppings within the cleared zone are counted and recorded. Separate records are kept for each section of each transect.

Recorders should ensure that they are able confidently to distinguish between the droppings of sheep, where these are present in the areas through which the transects pass, and those of rabbits and deer.

Information about rabbit and/or deer control at the site, or on parts of the site, by fencing, shooting, etc, should accompany the data, together with information on the prevalence of disease.

Author   *J.C. Coulson*

FROG SPAWN

Aims

To record the phenology of the spawning of the common frog in selected ponds and ditches by assessing the number of egg masses, as an indicator of the 'health' of frog populations.
To identify any significant changes in those populations and to measure some physical characteristics of the ponds which may affect breeding success

Rationale

The common frog (*Rana temporaria*) has been selected for inclusion in the ECN programme as an example of a ubiquitous predatory amphibian. Adult frogs feed mainly on insects, slugs and snails but also eat woodlice and the larvae of moths and butterflies. During the breeding season, frogs live in shallow ponds and ditches, but spend much time on land during the rest of the year. The frog population is therefore affected by conditions on land as well as in the shallow water bodies in which they breed. The changing status of frogs and other amphibians in Britain has been studied and reported (Cooke & Arnold 1982); populations have decreased, particularly during the 1960s (Cooke 1972), as a result of agricultural drainage, modification of breeding sites, and pollution by fertilizers and from other sources. In the last two decades there has been an increased awareness of several effects of the acidification of fresh waters by airborne pollutants and this has generated a parallel interest in the effects of increased acidity and concentrations of aluminium on larval amphibians (Cummins 1986; Beattie, Tyler-Jones & Baxter 1992). Numerous laboratory studies have demonstrated sublethal and lethal effects of acid conditions on embryonic and larval amphibians, and evidence of acid-related stress and mortality has also been found under field conditions (Cummins 1990).

It is difficult to monitor populations of adult frogs, but a rough approximation of frog colony size in a pond can be obtained by counting the number of egg clumps and multiplying by two; this assumes that each female produces one clump and that there is an even sex ratio (Cooke 1975). Frogs usually spawn in shallow water, up to about 15 cm depth, and the spawn masses and the water in which they rest are therefore usually reasonably accessible for monitoring purposes.

Method

Equipment

A pH meter suitable for use in the field, and a maximum/minimum thermometer are available from laboratory suppliers. Other equipment, such as a water depth measurement pole, is easily made.

Location

One or more shallow ponds or wet ditches will be selected, preferably those in which frogs are known to have bred in recent years and which are situated conveniently for frequent visits to record stages in development.

Sampling

Biological

The time at which frog breeding starts in the UK varies greatly; in some years it may begin during December in Cornwall whilst not starting until April at high altitudes in the Pennines and in Scotland. In a particular pond, however, annual variation in the date of spawning tends to be rather small. Recorders will check the pond(s) weekly from about 1 January, or from an earlier or later date where local knowledge is available, to ascertain and record the date on which male frogs congregate in the spawning areas and begin calling. Thereafter the pond(s) will, if possible, be visited daily until the first eggs have hatched. At sites where daily visits are impossible, because of time and distance constraints, visits should be made as often as possible, with a minimum of weekly visits, until the first eggs have hatched.

A record is made of the date on which spawn is first seen, and on subsequent daily or weekly visits the number of new spawn masses which have appeared since the last visit are recorded. Each new spawn mass is marked by carefully attaching a coloured thread to its edge, using a large bodkin needle. This will ensure that a mass is not counted twice. Newly deposited egg masses can be recognised because the eggs are packed tightly and the jelly capsules surrounding the embryos will not have expanded. Threads can be removed from egg masses which have been in the water for several days. The total surface area of water occupied by the spawn masses will be estimated in square metres. A percentage estimate of dead or obviously diseased eggs will be made. Where visits are made less frequently than daily, and in any case when the total number of spawn masses exceeds 100, it will be impossible to count new masses and only the total surface area of water occupied by the spawn masses will be estimated in square metres. The date on which embryos are first seen to have hatched from the eggs is recorded. Once embryos have started to hatch, weekly visits and recording (see next paragraph) are sufficient and will continue until newly metamorphosed frogs are seen leaving the pond, or for a period of 16 weeks from the time when spawn was first observed, whichever is the shorter.

BF Protocol

Physico-chemical

At the time when spawn is first seen on the pond a 250 ml water sample is taken from the spawning area, using a bottle which has been rinsed in pond water before filling. Analysis of this sample provides a means of characterising ponds across the network and an annual baseline against which any subsequent changes in chemical composition of individual ponds can be assessed. Conductivity and pH are measured on unfiltered water according to methods in the ECN Initial Water Handling (WH) Protocol, which also describes the method for filtering the sample. After filtering, the water is analysed for dissolved Na^+, K^+, Ca^{2+}, Mg^{2+}, Fe^{2+}, Al^{3+}, NH_4^+-N, Cl^-, NO_3^--N, SO_4^{2-}-S, PO_4^{3-}-P, alkalinity and dissolved organic carbon. If there is subsequently a high and unexplained mortality of spawn or tadpoles, a further water sample is to be taken for the chemical analyses specified above, as an aid to explaining the cause of such mortality.

pH, temperature and water depth are important factors affecting the breeding success of frogs and these will be measured and recorded at weekly intervals between the date of first spawning and the date when newly metamorphosed frogs are first seen leaving the pond.

pH will be measured weekly at each of three re-locatable, random positions immediately outside the spawning area and close to the edge of the pond at a water depth of 50 mm. Ideally, a field pH meter should be used, its electrode being placed in the water with its tip at a depth of 50 mm.

A maximum/minimum thermometer will be set up in an area of open water immediately outside the spawning area by attaching the thermometer to a float so that it is held horizontally in the water at a depth of 50 mm. A suitable float can be made from a sturdy plastic bottle, its buoyancy being adjusted by adding water to ensure that the thermometer is held at the correct depth. Alternatively the thermometer can be attached to the lower surface of an appropriately sized block of wood which floats with the thermometer in the shade. The float is attached to the bank with one or more lines so that is held in position. The water temperature at the time of first spawning is recorded and the thermometer is subsequently read and re-set each week.

Water depth is measured weekly by placing a pole, graduated in centimetres, in the water as near as practicable to the centre of the area occupied by spawn.

Authors   *R.C. Beattie, J.K. Adamson and J.M. Sykes*

BF Protocol

References

Beattie, R.C., Tyler-Jones, R. & Baxter, M.J. 1992. The effects of pH, aluminium concentration and temperature on the embryonic development of the European common frog, *Rana temporaria*. *Journal of Zoology*, **228**, 556–570.

Cooke, A.S. 1972. Indicators of recent changes in status in the British Isles of the frog (*Rana temporaria*) and the toad (*Bufo bufo*). *Journal of Zoology*, **167**, 161–178.

Cooke, A.S. 1975. Spawn site selection and colony size of the frog (*Rana temporaria*) and the toad (*Bufo bufo*). *Journal of Zoology*, **175**, 292–38.

Cooke, A.S. & Arnold, H.R. 1982. National changes in status of the commoner British amphibians and reptiles before 1974. *British Journal of Herpetology*, **6**, 206–207.

Cummins, C.P. 1986. Effects of aluminium and low pH on growth and development in *Rana temporaria* tadpoles. *Oecologia*, **69**, 248–252.

Cummins, C.P. 1990. Effects of acid waters on the survival of frogs (*Rana temporaria*). *Annual Report of the Institute of Terrestrial Ecology 1989–1990*, 38–40.

INITIAL WATER HANDLING (WH)

Aim *To measure the conductivity and pH of samples of precipitation, soil solution and river water and to filter these samples prior to chemical analysis*

Rationale From the moment water samples are gathered they begin to deteriorate as a result of chemical and microbiological processes. Three methods of slowing this deterioration are to be used in ECN – filtration, cold storage, and (for Al and Fe determinations only) acidification. Some of the activities detailed in this Protocol require laboratory training, experience, specialised equipment, and compliance with health and safety requirements (eg diluting concentrated acids); guidance or help from the chemists who will perform the water analyses may also be required. Further background and quality control procedures are given in the documents referred to below.

Method **Initial storage**

Samples must be placed in cold storage, at a temperature between 1°C and 4°C, if the interval between collection and the measurement of conductivity and pH is more than seven hours. They should be returned to cold storage if filtering is not completed on the same day as these measurements. If samples are to be sent to another site for initial handling, they should be sent on the day of collection by the speediest possible method. Samples in transit must be placed in a cool box with pre-frozen cool blocks.

Conductivity measurement

Conductivity must be measured within 36 hours of collection on an unfiltered subsample at a temperature of 25°C according to the method given by HMSO (1978). Results should be expressed to one decimal place (0.1 μS cm^{-1}).

pH measurement

pH must be measured within 36 hours of collection on an unfiltered subsample. The same subsample can be used both for conductivity and for pH measurement, but conductivity should be measured first and the subsample should not be returned to the main sample after measurement. The method to be used is given in HMSO (1988) which implies the use of separate glass and reference electrodes. Results should be expressed to two decimal places (0.01 pH unit).

Filtering

Filtering must take place within 60 hours of collection. The method to be used is given in Appendix II. Filtering details should be recorded, using the unique sample codes described in the Protocols (also described in Appendix I of this Protocol). After filtration the samples are to be analysed for dissolved Na^+, K^+, Ca^{2+}, Mg^{2+}, Fe^{2+}, Al^{3+}, NH_4^+ -N, Cl^-, NO_3^- -N, SO_4^{2-} -S, PO_4^{3-} -P, alkalinity and dissolved organic carbon (DOC). DOC is not mandatory for precipitation water and can be omitted. Separate documentation is provided for chemical analysis (see AG Protocol) in which the determinands are placed in an order of priority for those occasions on which sample volume is insufficient to allow the measurement of all determinands.

Storage prior to chemical analysis

Storage temperature should be between 1°C and 4°C. Analysis should be completed preferably within 16 days of collection but definitely within 28 days. If samples have to be sent to another location for analysis, the period when samples are out of cold storage should be minimised, for example by specifying next day delivery and not despatching on a Friday.

Washing bottles

The method for washing bottles is given in Appendix II.

Author   *J.K. Adamson*

References   HMSO. 1978. *The measurement of electrical conductivity and the laboratory determination of the pH value of natural, treated and waste waters.* London: HMSO.

HMSO. 1988. *The determination of pH in low ionic strength waters.* London: HMSO.

Appendix I   Filtering

Materials

- Filter funnel assembly – most modern filter assemblies are suitable, eg the Millipore Sterifil which holds 47 mm diameter filters (Millipore catalogue no. XX11 047 00)
- Membrane filters **(only)** – Whatman, type WCN, pore size 0.45 mm, obtainable, for example, from BDH, catalogue no. (to fit the above filtering system) 402/0723/04
- Glass fibre filters – Whatman, type GF/C, obtainable, for example, from BDH, catalogue no. (to fit the above filtering system) 234/0856/11
- Vacuum pump – a pump capable of maintaining a vacuum of approximately 0.5 bar

Procedure

All parts of the filter funnel assembly must be thoroughly rinsed with deionised or distilled water before the first sample and between subsequent samples. The filters and the surfaces of the filter assembly which come into contact with the water sample must not be touched by hand. Filters should be moved with forceps and each filter should be used for only one sample. Waters originating from soil solution (SS) and rivers (WQ) may need pre-filtering through a glass fibre filter but these filters require washing with 250 ml of distilled water before use. The filtered water should be poured from the filter assembly into a clean, dry polypropylene bottle and not returned to the bottle used for the unfiltered water. A bottle of 250 ml volume is likely to be appropriate, but the laboratory performing the analyses should be consulted. If the volume of a sample exceeds 100 ml, a subsample of 20 ml should be transferred from each sample to a vial for determination of Al and Fe, and acidified with 20 ml HCl.

All containers should be labelled with the full ECN sample code (the ECN measurement code, the ECN site number and the local sample code) and the date of collection, which together gives each sample a unique reference, eg SS-04-B1S 06-Jan-1993. This unique reference should appear with the final data submitted for inclusion in the ECN database.

Suppliers

BDH (Head Office)
Merck House
Poole   *Tel: 01202 664778*
Dorset BH15 1TD, UK   *Fax: 01202 666541*

Millipore (UK) Ltd
The Boulevard
Blackmoor Lane
Watford   *Tel: 01923 816375*
Hertfordshire WD1 8YW, UK *Fax: 01923 818297*

WH Protocol

Appendix II   Bottle and vial washing

Procedure

Containers must be washed in a laboratory cleaning agent before being used for the first time, and subsequently at approximately six-monthly intervals, or if subjected to high levels of soiling (eg contamination of rainfall by bird droppings). The cleaning agent should be free of phosphate and hypochlorite. If a laboratory washing machine is available, Decomatic (not Dri-Decon) is suitable, and, if no machine is available, over-night soaking in Decon-90 is suitable. Subsequently, containers will be rinsed four times in tap water and three times in deionised or distilled water. At other times, after use, containers will be rinsed three times in distilled water and retained for use with the same sampler on subsequent occasions.

After washing, containers must be shaken to remove drops of distilled water, dried in warm air in a dust-free environment, and re-capped immediately.

Suppliers

Decon products can be obtained from a number of suppliers, including BDH (see Appendix I).

ANALYTICAL GUIDELINES FOR WATER SAMPLES

Aim   *To provide guidelines for the analysis of water samples from ECN terrestrial sites*

Rationale   Whilst each laboratory has full responsibility for managing its own analytical resources, participating laboratories have an obligation to conduct analyses under controlled conditions, and to provide documentation of methodology, information about significant changes in methodology, and information about validation procedures.

Procedures will be subject to annual review by an Analytical Working Group with representatives from each laboratory.

Method   Appropriate methods are outlined below. Detection limits indicate targets for laboratories for their chosen method.

Approved techniques

Alternative approved techniques for each determinand are given in the following Table (superscripts refer to Notes section below).

pH	[1]HMSO (1988) ('Blue book')		
Conductivity	[2]HMSO (1978) ('Blue book')		
Na^+	FES/AAS	ICP/OES	IC
K^+	FES/AAS	ICP/OES	IC
Ca^{2+}	AAS	ICP/OES	IC
Mg^{2+}	AAS	ICP/OES	
Fe^{2+}	AAS	ICP/OES	
Al^{3+}	AAS	ICP/OES	[4]Col-PCV
NH_4^+ N	[4]Col-indophenol blue		
Cl^-	IC	[4]Col-Hg/thiocyanate	
NO_3^-N	IC	([4]Col-Greiss/Illosvay)	
SO_4^{2-} S	IC	(ICP/OES)	
HCO_3^-	Alkalinity. [3]HMSO (1981) ('Blue book')		
PO_4^{3-} P	[4]Col-molybdenum blue		
DOC	Combustion	Colorimetry	
Total N	Kjeldahl/indophenol blue	Persulphate/NO_3	

Note: Techniques shown in brackets do not quantify directly the species of interest, but may be suitable following comparative tests. For pH and conductivity measurements, please see also the Initial Water Handling (WH) Protocol

[1]**HMSO.** 1988. *The determination of pH in low ionic strength waters.* London: HMSO.

[2]**HMSO.** 1978. *The measurement of electrical conductivity and the laboratory determination of the pH value of natural, treated and waste waters.* London: HMSO.

[3]**HMSO.** 1981. *The determination of alkalinity and acidity in water.* London: HMSO. (Method B)

[4]Col - colorimetry using continuous flow or flow injection analysis

Reference techniques

Changes and developments in analytical technology will inevitably occur during the lifetime of the ECN programme, and it is important to have a series of reference methods with which to compare and to assess alternative or innovative analytical methodologies. The methods listed below are the most suitable for use as primary references.

	Results expressed as	Species of interest	Reference technique	Ultimate detection limit
pH	pH		HMSO (1988)	
Conductivity	$\mu S\ cm^{-1}$		HMSO (1978)	1
Na^+	$mg\ l^{-1}$		AAS	0.02
K^+	$mg\ l^{-1}$		AAS	0.01
Ca^{2+}	$mg\ l^{-1}$		AAS	0.02
Mg^{2+}	$mg\ l^{-1}$		AAS	0.02
Fe^{2+}	$mg\ l^{-1}$		AAS	0.1
Al^{3+}	$mg\ l^{-1}$		AAS	0.1
$NH_4^+\ N$	$mg\ l^{-1}\ N$		Indophenol blue	0.1
Cl^-	$mg\ l^{-1}$		IC	0.5
NO_3^-	$mg\ l^{-1}\ N$	Nitrate	IC	0.01
SO_4^{2-}	$mg\ l^{-1}\ S$	Sulphate	IC	0.1
Alkalinity	$mg\ l^{-1}\ CaCO_3$		HMSO (1981)	1
PO_4^{3-}	$mg\ l^{-1}\ P$	o-phosphate	Molybdenum blue	0.005
DOC	$mg\ l^{-1}\ C$		Combustion	0.1
Total N	$mg\ l^{-1}\ N$		Kjeldahl	0.1

Notes: HMSO references are given on previous page

For conductivity and pH measurement, please refer also to the Initial Water Handling (WH) Protocol. For pH measurement, a procedure should be used which is capable of quantifying low ionic strength solutions, with a separate glass and a recommended reference electrode.

Laboratories should report data on the fraction specified (eg SO_4^{2-} by ion chromatography). If there is a strong preference by the laboratory to use a technique which quantifies a slightly different fraction (eg S by ICP/OES), then there is an obligation on the laboratory to monitor the two alternative techniques for the ECN samples in order to provide data to confirm the absence of bias.

The analytical ranges in operation will differ between laboratories, producing different working detection limits. However, when the solution concentration is in the region of the working detection limit of the particular analytical system, there is a need to consider whether the quoted values at this detection limit provide data suitable for the purpose of the study, or alternatively whether the samples should be re-analysed using a lower analytical range. The defined limits provide a uniform guideline to be followed by each laboratory.

AG Protocol

Priority listing of determinands

Sample volumes for soil solution and precipitation waters will be limited in periods of low rainfall. The following priority listing (where 1 indicates the highest priority) provides a working guideline to assist site operators and analysts in making decisions on handling and analytical options when sample volume is limiting.

1. pH, conductivity
2. Anions – NO_3^-, SO_4^{2-}, then PO_4^{3-}, Cl^-
3. Cations – Ca^{2+}, Mg^{2+}, then K^+, Na^+
4. NH_4^+ N
5. DOC
6. Cations requiring separate acidified portion – Fe^{2+}, Al^{3+}
7. Total-N, alkalinity

Please refer to the Initial Water Handling (WH) Protocol for information on solution handling, filtration, pH and acidification of samples.

Providing details of methodology

Details of analytical methods used by a laboratory should be kept at the laboratory for its operators, and summary information provided for the ECN database. This information will reside, one data record for each determinand for each ECN site, in the meta database, linked to the data by site, core measurement, determinand and date, to indicate the methodology in operation and its precision. The format is illustrated below, using nitrate as an example.

Performance characteristics of the method

Laboratory / Site	Merlewood / Moor House
Substance determined	Nitrate
Basis of the method	Chemically suppressed ion chromatography
Types of sample	Rainwater (PC), stream water (WC), soil solution (SS)
Typical concentrations	PC: 0, WC: 0.50, SS: 3.2
Volume for analysis	10 ml
Calibration range	0.01 to 10 mg l^{-1} – slight deviation from linearity corrected for by using 3rd-order regression
Method of measurement	Peak area using integration / data system
Results reported	3 significant figures as N (mg l^{-1})
Detection limit	0.010 mg l^{-1}
Within batch standard deviation (mid-range) *	2% rsd
Interferences	None
Internal QC measure	CUSUM quality control chart
Accuracy measure	AQUACHECK

Notes: The detection limit is defined as 4.65 within-batch standard deviation* of the blank or a solution with a concentration close to the blank when no signal is detectable from the blank (n=10)
*A within-batch standard deviation in excess of 5% is unlikely to be acceptable

AG Protocol

When any details change, a new record will be added to the database with a date 'stamp'. Aspects of the analysis such as instrument maintenance, calibration, drift, and training of staff will be controlled by the laboratory. The above information provides a record of changes in methodology and an assessment of 'suitability for the purpose' of the data.

Validation

Analytical data validation will be maintained through the application of approved techniques, adoption of uniform detection limits, and internal quality control, with an obligation to participate in regular interlaboratory analysis, eg AQUACHECK. The two main measures of quality are accuracy and precision.

Accuracy

Accuracy describes closeness to the true value, but in practical terms it reflects the agreement of values amongst the wider analytical community. Agreement can be established through participation in a control sample scheme, such as AQUACHECK. However, the operational frequency of such schemes is necessarily less than that of the sampling events and it will therefore be necessary to record all accuracy check results to confirm conformity, as well as recording any correction procedures which may be applied.

Example of an accuracy record

Element	SO_4^{2-}
Fraction measured	Dissolved
Results reported as (eg SO_4 as S)	S
Date of analysis	23-Jan-1993
Method	Ion chromatography
Laboratory	ITE Merlewood
Scheme	AQUACHECK
Within error threshold	Y
Result	1.23
Mean	1.34
Percentage relative error	+8.9%
Corrective measures	None

Individual laboratories should send all accuracy results relevant to ECN samples, in the categories shown above, to the ECN Database Manager in machine-readable form.

Precision

Precision control will be ensured by the use of synthetic solutions as quality control (QC) samples for assessment of possible batch bias. QC data should be reported to Site Managers with each set of sample results. Laboratories should not submit data if the QC values indicate that the procedures are functioning unacceptably.

Author   *A.P. Rowland*

Chapter 3   ECN DATA REPORTING

Introduction

Section 1 of this Chapter lists the measurement variables and results generated by each ECN Protocol. An overview of procedures for data handling and formats for transfer of data to the ECN Central Co-ordination Unit (CCU) in machine-readable form is given in Section 2. Documentation giving detailed specifications for data transfer for each individual Protocol are not provided here but may be obtained from the ECN CCU on request.

Each sample or sampling occasion and associated measurements are uniquely identified by:

- Core Measurement Code, eg PC
  (see Appendix I for a list of codes)
- Site Identification Code, eg 04
  (see Appendix I for a list of current site IDs)
- Internal Location Code, eg 01
  Each site allocates its own code to replicate sample locations (eg different surface water collection sites)
- Sampling Date(/time), eg 10-Mar-1996
  Date on which sample was collected or data were recorded. This will include a time element as part of the unique identifier where sampling is more frequent than daily (eg core measurements MA,WD)

Standard recording forms for use in the field and/or laboratory can be obtained from the CCU; some example forms have been provided in Appendix II. The use of computerised forms and mapping in the field is encouraged where sites have access to field computers, provided that output can be translated into ECN standard formats; the development of these systems will be monitored for future ECN standards in computerised data capture.

Any number of pre-defined quality codes (see Section 2.4 and Appendix III) may be associated with each sample or recording occasion. The forms list the quality codes most appropriate to their respective ECN core measurement. In addition, data may be accompanied by free-format text descriptions of problems which might affect the quality of the data.

All data recorded should be accompanied by the name of the person responsible for sampling/recording and also sample identification where appropriate. Information provided during the first year of monitoring should include grid references of the locations of point-based surveys and maps showing the locations of transects or area-based surveys. These will be incorporated into the ECN geographical information system (GIS).

The measurement variables listed in Section 1 are those required for an individual instrument or sampling location at a single site. Note that the Core Measurement Code, Site ID, and Internal Location Code explained above are required to identify every set of data, but are not repeated in the listings below to save space. Superscripts in the following Tables refer to the Notes section associated with each core measurement.

1. Specification of results and recording conventions

1.1 Core measurement: meteorology – automatic weather station (MA Protocol)

The following variables are recorded hourly for each automatic weather station and automatically logged.

Variable	Units	Precision of recording
Year		
Day number	Julian within year	
Hour	GMT 24-h clock	1 h
Solar radiation (average)	W m^{-2}	1
Net radiation (average)	W m^{-2}	1
Wet bulb temperature (average)	°C	0.1
Dry bulb temperature (average)	°C	0.1
Wind speed (average)	m s^{-1}	0.1
Wind direction (average)	degrees	1
Rainfall (total)	mm	0.1
Albedo – ground (average)	W m^{-2}	1
Soil temperature at 10 cm (average)	°C	0.1
Soil temperature at 30 cm (average)	°C	0.1
Surface wetness (total time wetness <0.8)	min	1
Soil water potential (average)	bars	0.01

1.2 Core measurement: meteorology – manual (MM Protocol)

The following variables are recorded daily at 0900 GMT.

Variable	Units	Precision of recording
Recording (Sampling) date		
Recording (Sampling) time	GMT 24-h clock	1 min
Dry bulb temperature	°C	0.1
Wet bulb temperature	°C	0.1
Maximum temperature	°C	0.1
Minimum temperature	°C	0.1
Grass minimum temperature	°C	0.1
Soil temperature 30 cm	°C	0.1
Soil temperature 100 cm	°C	0.1
Rainfall (total)	mm	0.1
Wind run (total)	km	1

Recording forms

The standard British Meteorological Office recording form 3208B should be used for recording in the field, using the instruction booklet 3100A. Recorders should also refer to the Meteorological Office *Handbook* (1982).

1.3 Core measurement: atmospheric chemistry (AC Protocol)

The following variables are recorded fortnightly for each of three experimental and three blank diffusion tubes.

Variable	Units	Precision of recording
Setting out date		
Setting out time	GMT 24-h clock	1 min
Sampling date		
Sampling time	GMT 24-h clock	1 min
Tube code	2 character code[1]	
Weight NO_2^-	µg	3 significant figures

Recording forms

A standard field recording form is available from the CCU. An example is provided in Appendix II.

Note

1 Experimental tubes should be coded E1, E2, E3, blank tubes B1, B2, B3.

1.4 Core measurement: precipitation chemistry (PC Protocol)

The following variables are recorded weekly.

Variable	Units	Precision of recording
Setting out[1] date		
Setting out time	GMT 24-h clock	1 min
Sampling date		
Sampling time	GMT 24-h clock	1 min
Volume	ml	1
pH	pH scale	0.1
Conductivity	$\mu S\ cm^{-1}$	0.1
Alkalinity	$mg\ l^{-1}$	3 significant figures
Na^+	$mg\ l^{-1}$	3 significant figures
K^+	$mg\ l^{-1}$	3 significant figures
Ca^{2+}	$mg\ l^{-1}$	3 significant figures
Mg^{2+}	$mg\ l^{-1}$	3 significant figures
Fe^{2+}	$mg\ l^{-1}$	3 significant figures
Al^{3+}	$mg\ l^{-1}$	3 significant figures
$PO_4^{3-}-P$	$mg\ l^{-1}$	3 significant figures
NH_4^+-N	$mg\ l^{-1}$	3 significant figures
Cl^-	$mg\ l^{-1}$	3 significant figures
NO_3^--N	$mg\ l^{-1}$	3 significant figures
$SO_4^{2-}-S$	$mg\ l^{-1}$	3 significant figures

Recording forms
A standard field recording form is available from the CCU.

Note
1 Date/time bottle last emptied and set out

1.5 Core measurement: surface water discharge (WD Protocol)

The following variables are recorded automatically at 15 min intervals.

Variable	Units	Precision of recording
Recording (Sampling) date		
Recording (Sampling) time	GMT 24-h clock	1 min
Stage (average)	m	0.001
Discharge (average)	$m^3 s^{-1}$ (cumecs)	0.001

1.6 Core measurement: surface water chemistry (WC Protocol)

The following variables are recorded from weekly samplings.

Variable	Units	Precision of recording
Sampling date		
Sampling time	GMT 24-h clock	1 min
pH	pH scale	0.1
Conductivity	$\mu S\ cm^{-1}$	0.1
Alkalinity	$mg\ l^{-1}$	3 significant figures
Na^+	$mg\ l^{-1}$	3 significant figures
K^+	$mg\ l^{-1}$	3 significant figures
Ca^{2+}	$mg\ l^{-1}$	3 significant figures
Mg^{2+}	$mg\ l^{-1}$	3 significant figures
Fe^{2+}	$mg\ l^{-1}$	3 significant figures
Al^{3+}	$mg\ l^{-1}$	3 significant figures
$PO_4^{3-}-P$	$mg\ l^{-1}$	3 significant figures
NH_4^+-N	$mg\ l^{-1}$	3 significant figures
Cl^-	$mg\ l^{-1}$	3 significant figures
NO_3-N	$mg\ l^{-1}$	3 significant figures
$SO_4^{2-}-S$	$mg\ l^{-1}$	3 significant figures
Dissolved organic carbon	$mg\ l^{-1}$	3 significant figures

Recording forms
A standard field recording form is available from the CCU. An example is provided in Appendix II.

1.7 Core measurement: soils (S Protocol)

Soils - baseline recording (SB)
The following information is recorded in the first year of monitoring:
- Soil map at 1:10 000 scale (or 1:25 000 for larger sites)
- Soil typologies at 50 m intervals around TSS and 25 m intervals within TSS from auger borings

Soils fine-grain recording (SF)

The following variables are recorded for the initial and five-yearly samplings.

Variable	Units	Precision of recording
Sampling date		
Block code	1 character code[1]	
For each horizon (by block)		
Horizon code	character code[2]	
Horizon lower average depth	cm	1
Horizon average thickness	cm	1
For each depth band (by block)		
Depth code	numeric code[3]	
Then, for each depth and each horizon sample		
Soil moisture	%	0.1
pH	pH scale	0.1
Exchangeable		
Acidity	$mmol_c\ kg^{-1}$	0.1
Sodium	$mmol_c\ kg^{-1}$	0.01
Potassium	$mmol_c\ kg^{-1}$	0.01
Calcium	$mmol_c\ kg^{-1}$	0.01
Magnesium	$mmol_c\ kg^{-1}$	0.01
Manganese	$mmol_c\ kg^{-1}$	0.01
Aluminium (assume all Al is Al_3^+)	$mmol_c\ kg^{-1}$	0.01
Total		
Nitrogen	%	0.01
Phosphorus	$mg\ kg^{-1}$	10
Sulphur	$mg\ kg^{-1}$	0.01
Organic carbon	%	0.1
Inorganic carbon	%	0.1

Soils coarse-grain recording (SC)

The following variables are recorded for the initial and 20-yearly samplings for the each soil block.

Variable	Units	Precision of recording
Sampling date		
Block code	1 character code[1]	
Cell code	1 character code[4]	
Altitude	m	1
Slope	degrees	1
Aspect	degrees	1
Land use	numeric codes[5]	
Decomposition of peats	character code[6]	

For each horizon

Horizon code	character code[2]	
Horizon lower average depth	cm	1
Horizon average thickness	cm	1
Description	See note 7	
Particle size analysis	% in each size class[8]	1
Soil mineralogy	% in each mineral class[9]	1
Soil water release	% volume held at different tensions[10]	1
Bulk density	kg m^{-3}	1

For each depth band

Depth code	numeric code[3]	

Then, for each depth or horizon sample

Soil moisture	%	0.1
pH	pH scale	0.1
Exchangeable		
Acidity	mmol$_c$ kg^{-1}	0.1
Sodium	mmol$_c$ kg^{-1}	0.01
Potassium	mmol$_c$ kg^{-1}	0.01
Calcium	mmol$_c$ kg^{-1}	0.01
Magnesium	mmol$_c$ kg^{-1}	0.01
Manganese	mmol$_c$ kg^{-1}	0.01
Aluminium (assume all Al is Al$_3^+$)	mmol$_c$ kg^{-1}	0.01
Total		
Nitrogen	%	0.01
Phosphorus	mg kg^{-1}	10
Sulphur	mg kg^{-1}	0.01
Organic carbon	%	0.1
Inorganic carbon	%	0.1
Lead	mg kg^{-1}	0.01
Zinc	mg kg^{-1}	0.01
Cadmium	mg kg^{-1}	0.01
Copper	mg kg^{-1}	0.01
Mercury	mg kg^{-1}	0.01
Cobalt	mg kg^{-1}	0.01
Molybdenum	mg kg^{-1}	0.01
Arsenic	mg kg^{-1}	0.01
Chromium	mg kg^{-1}	0.01
Nickel	mg kg^{-1}	0.01
Extractable		
Iron	%	1
Aluminium	%	1
Phosphorus	mg kg^{-1}	1

Recording forms

The forms supplied in Hodgson (1974) should be used for field survey.

Notes

1. The six soil blocks are coded A–F. Refer to the Protocol for a

description of the sampling layout.

2. Horizons should be coded according to the notation described in Hodgson (1974). Codes must be unique within each profile; horizons qualifying for the same letter notation and occurring in vertical sequence are denoted by numerals placed after the letter designation (eg Ah1,Ah2).

3. Depth bands should be coded as follows:

0–5 cm	1	20–40 cm	5
5–10 cm	2	40–60 cm	6
10–20 cm	3	60–80 cm	7
20–30 cm	4 *	80–100 cm	8
		100–120 cm	9

* Bands 1–4 are used for five-yearly sampling (SF). Band 4 is excluded from 20-year sampling (SC), being replaced by band 5.

4. Cells for 20-year sampling are coded from A to P. Refer to the Protocol for a description of sampling layout.

5. Land use should be described using the numeric codes given in Hodgson (1974); these codes are also given here under Section 1.9, note 5, page 167.

6. Decomposition of peats: Table 3 is taken from Avery (1980). Please use the decomposition codes H1–H10 given in the Table.

7. Horizon descriptions should follow the procedures given in Hodgson (1974).

8. Particle size classes are as follows:

Clay (<2 μm)	1
Fine silt (2–63 μm)	2
Coarse silt (63–106 μm)	3
Fine sand (106–212 μm)	4
Med sand (212–600 μm)	5
Coarse sand (600 μm–2 mm)	6

9. Soil mineral categories are as follows:

Sand   Quartz
Potassium feldspar
Plagioclase feldspar
Carbonate minerals
Ferro-magnesium minerals:
     garnets, olivines, calcium and magnesium amphiboles, sodium-rich amphiboles, pyroxenes, biotite mica, muscovite, chlorite, epidotes

Silt   Quartz
Potassium feldspar
Plagioclase feldspar
Carbonate minerals
Ferro-magnesium minerals

Clay   Smectite
Kaolinite
Illite
Chlorite
Inter-stratified minerals
Iron-oxide minerals
Aluminium hydroxide minerals

Table 3. Modified version of the Von Post scale for assessing the degree of decomposition of peat (source: Avery 1980)

	Nature of liquid expressed on squeezing	Proportion of peat extruded between fingers	Nature of plant residues	Description
H1	Clear, colourless	None	Plant structure unaltered; fibrous, elastic	Undecomposed
H2	Almost clear, yellow-brown	None	Plant structure distinct; almost unaltered	Almost undecomposed
H3	Slightly turbid, brown	None	Plant structure distinct; most remains easily identifiable	Very weakly decomposed
H4	Strongly turbid, brown	None	Plant structure distinct; most remains identifiable	Weakly decomposed
H5	Strongly turbid, contains a little peat in suspension	Very little	Plant structure clear, but becoming indistinct; most remains difficult to identify	Moderately decomposed
H6	Muddy, much peat in suspension	One-third	Plant structure indistinct, but clearer in the squeezed residue than in the undisturbed peat; most remains unidentifiable	Well decomposed
H7	Strongly muddy	One-half	Plant structure indistinct but recognisable; few remains identifiable	Strongly decomposed
H8	Thick mud, little free water	Two-thirds	Plant structure very indistinct; only resistant remains such as root fibres and wood identifiable	Very strongly decomposed
H9	No free water	Nearly all	Plant structure almost unrecognisable; practically no identifiable remains	Almost completely decomposed
H10	No free water	All	Plant structure unrecognisable; completely amorphous	Completely decomposed

(In this field test a sample of wet peat is squeezed in the closed hand and the colour of the liquid that is expressed between the fingers, the proportion of the original sample that is extruded, and the nature of the plant residues are observed)

10. Water release tension categories are as follows:

50 millibars	1
100 millibars	2
400 millibars	3
2 bars	4
15 bars	5

1.8 Core measurement: soil solution chemistry (SS Protocol)

The following variables are recorded from fortnightly samplings for each of 12 soil solution samplers.

Variable	Units	Precision of recording
Setting out[1] date		
Setting out time	GMT 24-h clock	1 min
Sampling date		
Sampling time	GMT 24-h clock	1 min
Sampler code	3-character code (see SS Protocol, page 77)	
Volume	ml	10
Vacuum	bars	0.01
pH	pH scale	0.1
Conductivity	$\mu S\ cm^{-1}$	0.1
Alkalinity	$mg\ l^{-1}$	3 significant figures
Na^+	$mg\ l^{-1}$	3 significant figures
K^+	$mg\ l^{-1}$	3 significant figures
Ca^{2+}	$mg\ l^{-1}$	3 significant figures
Mg^{2+}	$mg\ l^{-1}$	3 significant figures
Fe^{2+}	$mg\ l^{-1}$	3 significant figures
Al^{3+}	$mg\ l^{-1}$	3 significant figures
$PO_4^{3-}-P$	$mg\ l^{-1}$	3 significant figures
NH_4^+-N	$mg\ l^{-1}$	3 significant figures
Cl^-	$mg\ l^{-1}$	3 significant figures
NO_3^--N	$mg\ l^{-1}$	3 significant figures
$SO_4^{2-}-S$	$mg\ l^{-1}$	3 significant figures
Dissolved organic carbon	$mg\ l^{-1}$	3 significant figures

Recording forms
A standard field recording form is available from the CCU.

Note
1   Date/time vacuum 'drawn-down'.

1.9 Core measurement: vegetation (V Protocol)

Vegetation baseline recording (VB)
Initial survey in first year of monitoring for approximately 400 grid plots and up to 100 'infill' plots over the whole ECN site

Variable	Units	Precision of recording
For each plot		
Sampling date		
Plot position ID	numeric code[1]	
Plot position type	character code: (G=Grid, I=Infill)	
Grid reference	UK National Grid (m)	1
Altitude	m	10
Plot type	character code	

162

	S=Standard 2 m	
	W=Woodland 10 m	
National Vegetation Classification	NVC category code[2]	
Species present	'VESPAN'[3] numeric codes	
Plot features present	character codes[4]	
	eg ditch, wall, bank)	
% plot unsurveyed	%	10
(eg if obstruction)		

Vegetation coarse-grain recording (VC)

Initially and again every nine years for 50 randomly selected plots from the original grid and infill positions.

Variable	Units	Precision of recording
For each plot		
Sampling date		
Plot position ID	numeric code[1]	
Land use	Hodgson[5] codes	
Slope	degrees	1
Aspect	degrees	1
Slope form	Hodgson[5] codes	
For each cell		
Cell ID	numeric code[6]	
Species present	'VESPAN'[3] numeric codes	
Cell features features present	character codes[4]	
(eg ditch, wall, bank)		

Vegetation woodland recording (VW)

Initially and again every nine years (except diameter at breast height (dbh) every three years) for those coarse-grain plots falling in scrub or woodland.

Variable	Units	Precision of recording
Sampling date		
Plot position ID		
Species present	'VESPAN'[3] numeric codes	
Dominance categories	character codes[7]	
For each of ten randomly selected trees		
Species	'VESPAN'[3] numeric codes	
Diameter at breast height (dbh)	cm	0.1
Height	m	0.5
Number of stems (eg if coppiced)	count	1
Distance of stem from centre		
of random cell	m	0.1

For each of ten randomly selected cells
Cell ID	numeric code[6]
Cell co-ordinates	local co-ordinates[6]
Species of seedling	'VESPAN'[3] numeric codes
Number of seedlings of each species	

Vegetation fine-grain recording (VF)

Initially and again every three years for at least two random plots in the TSS and two within each NVC type, from the original grid where possible.

Variable	Units	Precision of recording
For each plot		
Sampling date		
Plot position ID	numeric code[1]	
Land use	Hodgson[5] codes	
Slope	degrees	1
Aspect	degrees	1
Slope form	Hodgson[5] codes	
For each cell		
Cell ID	numeric code[6]	
Cell co-ordinates	local co-ordinates[6]	
Species present	'VESPAN'[3] numeric codes	
Plot features present	character codes[4]	
(eg ditch, wall, bank)		

Vegetation boundaries recording (VH)

Initially and again every three years for selected vegetation boundary plots.

Variable	Units	Precision of recording
For each transect plot		
Sampling date		
Plot position ID	numeric code[1]	
Grid reference	UK National Grid (m)	1
Altitude	m	10
Sketch map of transect profile		
For each transect cell		
Cell ID	numeric code[6]	
Species present	'VESPAN'[3] numeric codes	
Land use	Hodgson[5] codes	
Slope	degrees	1
Aspect	degrees	1
Slope form	Hodgson[5] codes	
Plot features present	character codes[4]	
(eg ditch, wall, bank)		

Species present	'VESPAN'[3] numeric codes

Vegetation: permanent grass (VP)

Four times per year.

Variable	Units	Precision of recording
For each exclusion cage		
Cage number	numeric code	
Dry matter yield	kg ha^{-1}	1

Vegetation: continuous cereal (VA)

Annually.

Variable	Units	Precision of recording
For each species recording plot (10 m x10 m)		
Sampling date		
Plot position ID	numeric code[1]	
For each cell		
Cell ID	numeric code[6]	
Cell co-ordinates	local co-ordinates[6]	
Species present	'VESPAN'3 numeric codes	
For each cereal plot (20–25 m length)		
Plot position ID	numeric code[1]	
Weight of grain	kg ha^{-1}	1
Dry matter yield of straw	kg ha^{-1}	1
For winter wheat crop		
Plot position ID	numeric code[1]	
Date of start of stem elongation (GS 31)		
Date flag leaf emerged (GS 39)		

Recording forms

Fifteen field recording forms have been designed for ECN vegetation monitoring, as listed below; these are available from the CCU. Example recording forms for vegetation baseline monitoring (VB) are provided in Appendix II.

1. Plot position and status information
2. Baseline (VB) species recording
3. Baseline (VB) descriptive information – plots
4. Coarse-grain (VC) species recording
5. Coarse-grain (VC) environmental information – plots
6. Coarse-grain (VC) descriptive information – cells
7   Fine-grain (VF) species recording
8. Fine-grain (VF) environmental information – plots
9. Fine-grain (VF) descriptive information – cells
10. Boundaries (VH) species recording

11. Boundaries (VH) environmental information – cells
12. Boundaries (VH) hedgerow species
13. Woodland (VW) species and dominance
14. Woodland (VW) tree diameter and height
15. Woodland (VW) seedlings

Notes

1. Numbering plot positions: All plot *positions* (for plots and transects) should be given a unique reference number within each ECN site. Different types of plots centred on and recorded at these positions will be referenced by a status code (eg VB, S for Baseline Standard plot)
2. National Vegetation Classification: see Rodwell, J.S. (1991).
3. Species naming and coding

   The recording forms supply the abbreviated names of the 200 most common species in GB (from a GB sample of 11 000 plots surveyed by the ITE Countryside Survey 1990 (Barr *et al.* 1993)). Species names conform to *Flora Europaea*, which should be used as the standard for ECN vegetation recording. Each species has an associated *unique* code number, used by the National Vegetation Classification system (see Malloch 1988 – available from: Institute of Environmental and Biological Sciences, University of Lancaster, Lancaster LA1 4YQ, UK).

   Note that, although the abbreviated names used on the forms are unique within the list of the 200 most common species, they are not necessarily unique for the complete list of all species. A list of the full names and codes of the 200 most common species is supplied in Appendix II with the example vegetation recording forms. A complete list of full names and codes of all species (*c* 4500 records) can be provided by the CCU.

   If a species is found which is not included in the list given on the form, it should be recorded in the 'Other species' section. The code number can be looked up in the complete species list and added later in the laboratory. For those genera whose species are considered difficult to distinguish, the genus name only has been provided on the form. If a surveyor is uncertain in distinguishing between species for a plant, then the 'Other species' section should be used to put down only the genus name.

   The (c), (s), and (g) codes which follow some species on the lists refer to canopy layer, shrub layer and ground layer respectively – different code numbers exist for the different layers. If you find a species which exists on the recording form list, but which is in a different layer, then use the 'Other species' section to record it. For example, a young *Betula pendula* seedling will have to be recorded in the 'Other species' section as 'Betu pendu (g)', as only the canopy layer – 'Betu pendu (c)' – is given in the list on the recording form. VESPAN codes are also used to record the presence of bare soil, bare rock, litter, dead wood and open water.

4. Feature codes

P=Path	S=Stream
W=Wall	H=Hedge

```
D=Ditch          N=Natural boundary
F=Fence          X=No feature
B=Bank
```

5. Hodgson slope form and land use codes and their descriptions are given in Hodgson (1974). They are summarised as follows.

Slope form

1	Convex
2	Straight (rectilinear)
3	Concave

Land use codes

- *Enclosed farmland*

1	Ley grassland	6	Root crops
2	Permanent or long-term grassland	7	Horticultural crops
3	Rough grazing	8	Fallow
4	Cereals	9	Other crops (specify)
5	Green crops	10	Orchard

- *Semi-natural vegetation, woodland and unenclosed farmland*

11	Deciduous woodland	16	Grassland
12	Coniferous woodland	17	Saltmarsh
13	Scrub	18	Fen, moor or bog
14	Lowland heath	19	Montane vegetation
15	Heather moor		

- *Land in other use*
20 Public park
21 Golf course
22 Other (specify)

6. Cells and points within plots
These should be given a local reference number as follows.
Coarse-grain cells (40 cm x 40 cm) within the 2 m x 2 m plots should be labelled from 1 to 25 as shown on the recording form, with 1 being the NW cell, 2 being the next cell to the east, and 25 being the SE cell. These correspond to the 25 numbered columns on the field form.
Fine-grain cells (40 cm x 40 cm) within 10 m x 10 m plots should be centred on randomly selected co-ordinate points (1–24 in X and Y, ignoring the edges of the plot). Each cell should be given a local reference number from 1 to 10 which will then correspond with the numbered columns of the respective recording form. The local co-ordinates referencing the centrepoints of each of the 10 cells within each plot should be provided for the ECN database.
Woodland points and cells within 10 m x 10 m plots should be . randomly selected, handled and numbered in the same way as the fine-grained cells described above. (Note that the diagrams on the fine-grain and woodland forms are purely illustrative and in no way indicate the location of the cells or points within the plot.)
Boundary transect cells should be labelled from –n to n,

corresponding with the columns of the recording form (cells −10 to +10 have been provided for, so far). Cell 1 should be the 'origin' cell and have one edge as near as possible to the middle of the linear feature or boundary. Cells can be added in future years as boundaries change, and cells labelled using the next positive or negative integer.

7. Woodland dominance categories

C = canopy dominant   U = suppressed   E = seedling
S = subdominant   H = shrub layer
I = intermediate   A = sapling

1.10 Core measurement: invertebrates – tipulids (IT Protocol)

The following variables are recorded twice yearly, in April and September.

Variable	Units	Precision of recording
Sampling date		
Core ID	character code (Cn)[1]	
Species code	BRC code[2]	
Species name	genus species	
Number found	count	1

Recording forms
Three forms are available from the CCU for up to three tipulid sampling sites within each ECN site.

Notes
1. Core/subplot IDs should be unique within each ECN site. For example, if there are three tipulid sampling locations, each taking 20 cores, the cores should be labelled from 1 to 60 (C1–C20 for Location 01, C21–C40 for Location 02, C41–C60 for Location 03).
2. The coding system should follow the standard currently used by the Biological Records Centre, ITE Monks Wood, Abbots Ripton, Huntingdon, Cambs PE17 2LS, UK.

1.11 Core measurement: invertebrates – moths (IM Protocol)

The following variables are recorded daily.

Variable	Units	Precision of recording
Date trap set		
Sampling date[1]		
Species code	RIS code[2]	
Species name	genus species	
Number caught	count	1

Recording forms
Please refer to the Rothamsted Insect Survey, Institute of Arable Crops Research, Rothamsted Experimental Station, Harpenden, Herts, UK.

Notes

1. The Sampling Date is the date on which the catch is collected following overnight recording. Any nights on which the trap was not operating, for example because of lamp failure, should also be reported, using the Sampling Date.
2. The coding system should follow the Heslop (1964) numbering system used by the Rothamsted Insect Survey.

1.12 Core measurement: invertebrates – butterflies (IB Protocol)

The following variables are recorded weekly from 1 April until 29 September:

Variable	Units	Precision of recording
For the transect		
Recording (Sampling) date		
Start time	BST 24-h clock	1 min
Temperature at end of recording	°C	0.1
Percentage sunshine	%	10
Wind speed at end of recording	Beaufort scale	
For each transect section		
Transect section number	numeric code (1–15)	
Species code	BMS code[1]	
Species name	common name	
Number seen	count	1
Habitat description	BMS method[2]	

Recording forms

A standard BMS recording form is available from the Butterfly Monitoring Scheme (BMS), organised from the Biological Records Centre, ITE Monks Wood, Abbots Ripton, Huntingdon, Cambs PE17 2LS, UK.

Notes

1. The coding system should follow the standard used by the BMS.
2. Habitat should be described as explained in the BMS handbook, quoted below.

   'A short description of the different habitat types in each section is useful for the transect records. It is also useful to have a short list of the plant species which are most abundant in each section. Particular attention should be given to butterfly food plants, eg nettles or violets, and popular nectar sources, such as thistles or teasels. The aim of these records is not to acquire quantitive information on the abundance of plants (which would be ideal but which is very time-consuming) but to help, in a general way, with the interpretation of results.'

1.13 Core measurement: invertebrates – spittle bugs (IS Protocol)

The following variables are recorded annually – nymphs in mid-June, adults in late August:

Variable	Units	Precision of recording
For nymph recording		
Sampling date		
Vegetation mixed or segregated	character code (M or S)	
Quadrat ID	character code (Qn)[1]	
Species code	BRC code[2]	
Species name	genus species	
Number of spittles	count	1
From random-throw exercise		
Number of spittles found (by species)	count	1
Total number of nymphs found (by species)	count	1
For P. spumarius adult morph recording		
Sampling date		
Morph	3-character code[3]	
Number of males	count	1
Number of females	count	1

Recording forms

Three types of form are available from the CCU:
1. Nymphs: quadrat sampling where the vegetation is mixed
2. Nymphs: quadrat sampling where the vegetation is segregated
3. Adults (sweep-netting).

Notes

1. Quadrats should be numbered from Q1 to Q20, or from Q1 to Q40, depending on whether the vegetation is mixed or separate. Numbering is unique within each ECN site.
2. The coding system should follow the standard currently used by the Biological Records Centre, ITE Monks Wood, Abbots Ripton, Huntingdon, Cambs PE17 2LS, UK.
3. Adult colour morphs of *P. spumarius* should be coded as follows:

POP	*populi*		PRA	*praeusta*
TRI	*trilineata*		LAT	*lateralis*
MAR	*marginella*		QUA	*quadrimaculata*
FLA	*flavicollis*		ALB	*albomaculata*
GIB	*gibba*		LOP	*leucopthalma*
LCE	*leucocephala*		TYP	*typica*
UUU	Unallocated morph			

1.14 Core measurement: invertebrates – ground predators (IG Protocol)

The following variables are recorded fortnightly from May until the end of October (13 trapping periods) for each transect of ten traps. Each transect of ten traps is regarded as a single sampling location.

Variable	Units	Precision of recording
Date traps set		
Collection (Sampling) date		
For all ground predator species		
Trap ID	character code (Tn)[1]	
Species code	BRC code[2]	
Species name	genus species	
Number caught	count	1
In addition, for harvestmen		
Species code of individual	BRC code[2]	
Species name of individual	genus species	
Gender of individual	M or F	
Femur length of individual	mm	0.1

Recording forms
Three types of form are available from the CCU:
1. Species recording form: one for each of the three transects (an example is provided in Appendix II)
2. Pitfall trap dates and quality information form (an example is provided in Appendix II)
3. Form for recording femur lengths for harvestmen.

Notes
1. Trap ID numbers should be unique within each ECN site, eg traps T1–T10 for transect 01, T11–T20 for transect 02, and T21–T30 for transect 03. For species recording (Form 1 above), separate rows of the Table should be used to record the different leg colour morphs and also gender of *Pterostichus madidus,* using the following codes after the species name:

FR	female red legs	MR	male red legs
FB	female black legs	MB	male black legs

2. The coding system should follow the standard currently used by the Biological Records Centre, ITE Monks Wood, Abbots Ripton, Huntingdon, Cambs PE17 2LS, UK.

1.15 Core measurement: vertebrates – birds (BI Protocol)

Common birds (BC)
Ten visits to map the location and behaviour of birds in a selected area are made per year, between mid-March and late June. The British Trust for Ornithology (BTO) Common Birds Census (CBC) method records habitat

British Trust for Ornithology (BTO) CBC habitat mapping instructions

(Taken from BTO Common Birds Census instructions)

Information on the nature of the habitat is an essential complement to the data you supply on the numbers and distribution of the territorial birds on your plot. It enables us to assess how representative is our index (by comparing the habitat of our plots with that of farmland or woodland as a whole), to compare the birds on plots of different habitat and, most importantly perhaps, to measure the effects on birds of specified changes in the environment.

If the habitat of your plot is subjected to major change, subsequent census results may form the basis of a detailed case study. We are likely to welcome the continuation of a census following such change, even where the changed area is substantially less attractive for birds, but please check with us first to ensure that results will be worthwhile.

The following items are needed annually to accompany each completed census:

(a) **a habitat map**. A full habitat map is essential in the first year, but in subsequent years it is necessary only to show changes from the previous year's map, and any special information which is relevant to that year (including field use on farmland plots). Details on compiling habitat maps are given below.

(b) **a completed habitat questionnaire**. Each observer will be sent a questionnaire before the start of the season, to be completed as fully as possible and returned with the maps. The content of the questionnaire may vary from year to year but for farmland will include field-use (cropping, management, farm chemicals used, etc), hedgerow management and other detailed aspects of habitat change. If there has been no change on the plot, whether farmland or woodland, this will be your opportunity to say so.

In addition, photographs of the plot are very helpful to the analyst, since they give an accurate impression of the habitat; they must be regarded as a complement to the habitat maps and questionnaires, not a substitute. Colour slides are particularly welcome. Please enclose them with a map showing the points from which the photographs were taken, and a note of the date.

Farmland habitat maps

In your first season, and in any subsequent season if you wish, please complete a full habitat map. This should be on one of the outline maps sent to you for the census and should describe the permanent skeleton of the plot – including any hedges, fences, ditches, tracks and lanes, farmsteads gardens, scrub, copses, permanent pasture, streams and standing water – together with a note of the field use in that season. Conventions are to mark hedgerows and wooded areas in green, and any streams or standing water in blue. Mapping should extend for 50–100 metres beyond the plot boundaries. The following details should be given:

(a) the plot boundaries, clearly marked
(b) contours, copied from the 6" or 2½" OS maps
(c) a six-figure grid reference for a point near the centre of the plot
(d) a description of each copse or block of woodland (see woodland section below)
(e) the structure of each hedge in terms of height, width, shape, main species of hedgerow shrubs and species and height of standard trees. The positions of standard trees should be marked with a cross
(f) position of any nestboxes
(g) any other details you think may affect the distribution of birds on your plot.

Estimate hedge width at the height at which the width is greatest; for hedges not recently trimmed it may be necessary to give ranges for height and width rather than single values.

A full habitat map will be welcome in any subsequent year of the census, and would be particularly useful following a period of habitat change, but the only requirement following the initial year is for a 'crops and changes' map. This should show:
(a) any changes in the habitat since the map for the preceding year, eg hedgerow losses, streams which have been dredged
(b) the cropping or field-use
(c) the hedgerows present in that year, marked with a green line
(d) the period of the season for which any standing water was present.

The 'crops and changes' map can be used to illustrate points you mention in your answers on the annual questionnaire. Please remember that unless you inform us of changes we might assume that the information on your previous habitat map is correct, so it is very important to keep up to date with recording habitat change.

Habitat information is best collected during the course of normal visits, but make a special visit if you wish. Notes made on the visit maps should be cancelled as they are copied to the habitat map.

Woodland habitat maps

As for farmland, a full habitat map is requested to accompany your first census. Please read the section on farmland habitat maps and mark all the features listed there, where relevant to the habitats present on your plot and in the surrounding 50–100 metre zone. In addition, the following specifically woodland features should be recorded:
(a) rides, clearings and glades
(b) boundaries between the major stand types, together with a brief description of each type.

Stand types can be recognised as blocks of woodland within which the tree and shrub species and the woodland structure are broadly uniform. Please provide the following details for each stand:

1. **Management type:** stands may vary in management (eg high forest, wood pasture, active coppice, derelict coppice). In coppiced woods,

the boundaries of different ages of coppice should be marked and the approximate date of cutting provided. Please inform us of any management activity on the annual questionnaire.

2. **Canopy or tree layer:** list the dominant species of trees and estimate by eye the approximate % cover for each tree species contributing more than 10% of the total ground cover. Also estimate the typical height of the dominant tree species: BTO staff can advise on methods if necessary.

3. **Shrub layer:** (1–5 m above ground): list the main species, their typical height and approximate % cover.

4. **Field layer and ground composition**: record the approximate % cover of grass, heather, herbs, bracken, bramble, rocks, etc.

Many plots contain only three or four different stand types which can be readily identified, and it will be rare to need more than seven or eight. A friendly botanist may be able to assist. If in difficulty consult the BTO. Please remember to keep us informed of any changes in habitat in subsequent years. Maps showing changes only would be welcome in addition to the completed questionnaire.

Note 2 — BTO species codes

Code	Species	Code	Species	Code	Species
AC	Arctic Skua	GE	Green Sandpiper	RL	Red-legged Partridge
AE	Arctic Tern	G.	Green Woodpecker	NK	Red-necked Phalarope
AV	Avocet	GR	Greenfinch	RH	Red-throated Diver
BY	Barnacle Goose	GK	Greenshank	LR	Redpoll
BO	Barn Owl	H.	Grey Heron	RK	Redshank
BR	Bearded Tit	GJ	Greylag Goose	RT	Redstart
BI	Bittern	P.	Grey Partridge	RE	Redwing
BK	Black Grouse	GL	Grey Wagtail	RB	Reed Bunting
BH	Black-headed Gull	GU	Guillemot	RW	Reed Warbler
BW	Black-tailed Godwit	HF	Hawfinch	RZ	Ring Ousel
BV	Black-throated Diver	HH	Hen Harrier	RI	Ring-necked Parakeet
BX	Black Redstart	HG	Herring Gull	RP	Ringed Plover
B.	Blackbird	HY	Hobby	R.	Robin
BC	Blackcap	HZ	Honey Buzzard	DV	Rock Dove
TY	Black Guillemot	HC	Hooded Crow	RC	Rock Pipit
BN	Black-necked Grebe	HP	Hoopoe	RO	Rook
BJ	Black Tern	HM	House Martin	RS	Roseate Tern
BU	Bluethroat	HS	House Sparrow	RY	Ruddy Duck
BT	BlueTit	JD	Jackdaw	RU	Ruff
BL	Brambling	J.	Jay	SM	Sand Martin
BF	Bullfinch	K.	Kestrel	TE	Sandwich Tern
BZ	Buzzard	KF	Kingfisher	VI	Savi's Warbler
C.	Carrion Crow	KI	Kittiwake	SQ	Scarlet Rosefinch
CG	Canada Goose	LM	Lady Amherst's Pheasant	SP	Scaup
CP	Capercaillie	LA	Lapland Bunting	CY	Scottish Crossbill
CW	Cetti's Warbler	L.	Lapwing	SW	Sedge Warbler
CH	Chaffinch	TL	Leach's Petrel	NS	Serin
CC	Chiffchaff	LB	Lesser B.b. Gull	SA	Shag
CF	Chough	LS	Lesser Sp. Woodpecker	SU	Shelduck
CL	Cirl Bunting	LW	Lesser Whitethroat	SX	Shorelark
CT	Coal Tit	LI	Linnet	SE	Short-eared Owl
CD	Collared Dove	LG	Little Grebe	SV	Shoveler
CM	Common Gull	LU	Little Gull	SK	Siskin
CS	Common Sandpiper	LO	LittleOwl	S.	Skylark
CX	Common Scoter	LP	Little Ringed Plover	SZ	Slavonian Grebe
CN	Common Tern	AF	Little Tern	SN	Snipe
CE	Corncrake	LE	Long-eared Owl	SB	Snow Bunting
CO	Coot	LT	Long-tailed Tit	ST	Song Thrush
CA	Cormorant	MG	Magpie	SH	Sparrowhawk
CB	Corn Bunting	MA	Mallard	AK	Spotted Crake
CI	Crested Tit	MN	Mandarin	SF	Spotted Flycatcher
CR	Crossbill	MX	Manx Shearwater	SG	Starling
CK	Cuckoo	MR	Marsh Harrier	SD	Stock Dove
CU	Curlew	MT	Marsh Tit	SC	Stonechat
DW	Dartford Warbler	MW	Marsh Warbler	TN	Stone-curlew
DI	Dipper	MP	Meadow Pipit	TM	Storm Petrel
DO	Dotterel	MU	Mediterranean Gull	SL	Swallow
DN	Dunlin	ML	Merlin	SI	Swift
D.	Dunnock	M.	Mistle Thrush	TO	Tawny Owl
EG	Egyptian Goose	MH	Moorhen	T.	Teal
E.	Eider	MO	Montagu's Harrier	TK	Temminck's Stint
FP	Feral Pigeon	MS	Mute Swan	TP	Tree Pipit
FF	Fieldfare	N.	Nightingale	TS	Tree Sparrow
FC	Firecrest	NJ	Nightjar	TC	Treecreeper
F.	Fulmar	NH	Nuthatch	TU	Tufted Duck
GA	Gadwall	OP	Osprey	TD	Turtle Dove
GX	Gannet	OC	Oystercatcher	TW	Twite
GW	Garden Warbler	PE	Peregrine	WA	Water Rail
GY	Garganey	PH	Pheasant	W.	Wheatear
GC	Goldcrest	PF	Pied Flycatcher	WM	Whimbrel
EA	Golden Eagle	PW	Pied Wagtail	WC	Whinchat
OL	Golden Oriole	PT	Pintail	WH	Whitethroat
GF	Golden Pheasant	PO	Pochard	WS	Whooper Swan
GP	Golden Plover	PM	Ptarmigan	WN	Wigeon
GN	Goldeneye	PU	Puffin	WT	Willow Tit
GO	Goldfinch	PS	Purple Sandpiper	WW	Willow Warbler
GD	Goosander	Q.	Quail	WO	Wood Warbler
GI	Goshawk	RN	Raven	WK	Woodcock
GH	Grasshopper Warbler	RA	Razorbill	WL	Woodlark
GB	Great B.b. Gull	RG	Red Grouse	WP	Woodpigeon
GG	Great Crested Grebe	ED	Red-backed Shrike	OD	Wood Sandpiper
ND	Great Northern Diver	RM	Red-breasted Merganser	WR	Wren
GS	Great Spotted Woodpecker	RQ	Red-crested Pochard	WY	Wryneck
NX	Great Skua	FV	Red-footed Falcon	YW	Yellow Wagtail
GT	Great Tit	KT	Red Kite	Y.	Yellowhammer

Note 3   BTO bird activity map symbols

BRITISH TRUST FOR ORNITHOLOGY (BTO):   BIRD ACTIVITY MAP SYMBOLS
(Sheet reproduced from BTO instructions for CBC recorders)

This standard list of conventions is designed for clear and unambiguous recording. Symbols can be combined where necessary. Additional activities of territorial significance, such as display or mating, should be noted using an appropriate clear abbreviation.

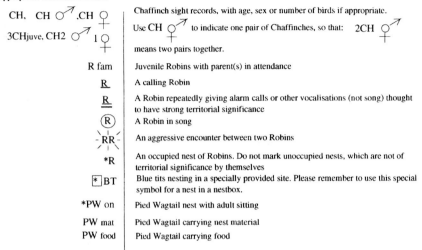

CH, CH ♂ .CH ♀ 3CHjuve, CH2 ♂ 1♀	Chaffinch sight records, with age, sex or number of birds if appropriate. Use CH ♀ to indicate one pair of Chaffinches, so that: 2CH ♂♀ means two pairs together.
R fam	Juvenile Robins with parent(s) in attendance
R	A calling Robin
R	A Robin repeatedly giving alarm calls or other vocalisations (not song) thought to have strong territorial significance
℞	A Robin in song
-RR-	An aggressive encounter between two Robins
*R	An occupied nest of Robins. Do not mark unoccupied nests, which are not of territorial significance by themselves
⊡BT	Blue tits nesting in a specially provided site. Please remember to use this special symbol for a nest in a nestbox.
*PW on	Pied Wagtail nest with adult sitting
PW mat	Pied Wagtail carrying nest material
PW food	Pied Wagtail carrying food

Movements of birds can be indicated by an arrow using the following conventions:

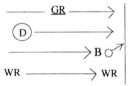

——— GR ———→	A calling Greenfinch flying over (seen only in flight)
(D)————→	A singing Dunnock, perched then flying away (not seen to land)
———→ B ♂	A male blackbird flying in and landing (first seen in flight)
WR ———→ WR	A Wren moving between two perches. The solid line indicates that it was **definitely the same bird**.

The following conventions indicate which registrations relate to different, and which to the same individual birds. Their proper use will be essential for the accurate assessment of clusters.

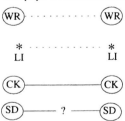

(WR)· · · · · ·(WR)	Two Wrens in song at the same time, i.e. definitely different birds. The dotted line indicates a simultaneous registration (or contemporary contact) and is of very great value in separating territories.
*LI · · · · · · *LI	Two Linnet nests occupied simultaneously, and thus belonging to different pairs. This is another example of the value of dotted lines. Only adjacent nests need to be marked in this way.
(CK)————(CK)	The solid line indicates that the registrations definately refer to the same bird.
(SD)—— ? ——(SD)	This question-marked solid line indicates that the registrations relate to probably the same bird. This convention is of particular use when your census route brings you back past an area already covered - it is possible to mark new positions of (probably the same) birds recorded before, without risk of double-recording. If you record birds without using the question-marked solid line, over-estimation of territories will result.
(WR) WR mat	No line joining the registrations - it will be assumed that the birds are probably different, but depending on the pattern of other registrations they may be treated as if only one bird was involved. (You may if you wish use a question-marked dotted line, indicating that the registations were almost certainly of different birds.)
C* C*	Where adjacent nests are marked without a line, it will often be assumed that they were in first and second broods, or a replacement nest following an earlier failure.

178

Note 4   BTO BBS county codes

Always fill in the county on recording and summary sheets using the four-letter code from the list below

England

Avon	GBAV
Bedford	GBBD
Berkshire	GBBK
Buckinghamshire	GBBC
Cambridgeshire	GBCA
Cheshire	GBCH
Cleveland	GBCV
Cornwall	GBCO
Cumbria	GBCU
Derbyshire	GBDB
Devon	GBDV
Dorset	GBDO
Durham	GBDU
Essex	GBES
Gloucestershire	GBGL
Hampshire (excl IoW)	GBHA
Hereford & Worcs	GBHF
Hertfordshire	GBHT
Humberside	GBHU
Isle of Wight	GBIW
Kent	GBKE
Lancashire	GBLA
Leicestershire	GBLE
Lincolnshire	GBLI
London (Greater)	GBLO
Manchester (Greater)	GBMA
Merseyside	GBME
Norfolk	GBNK
Northamptonshire	GBNH
Northumberland	GBNL
North Yorkshire	GBNY
Nottinghamshire	GBNT

Oxfordshire	GBOX
Shropshire	GBSA
Scilly Isles	GBSI
South Yorkshire	GBSY
Staffordshire	GBST
Somerset	GBSO
Suffolk	GBSK
Surrey	GBSR
Sussex (West & East)	GBSX
Tyne & Wear	GBTY
Warwickshire	GBWK
West Midlands	GBWM
West Yorkshire	GBWY
Wiltshire	GBWT

Isle of Man

Isle of Man	GBIM

Scotland

Borders Region	GBBR
Central Region	GBCR
Dumfries & Galloway Region	GBDR
Fair Isle	GBFI
Fife Region	GBFR
Grampian Region	GBGR
Highland Region	GBHR
Lothian Region	GBLR
Orkney	GBOR
Shetland	GBSH
Strathclyde Region	GBSC
Tayside Region	GBTR
Western Isles	GBWI

Wales

Anglesey	GBAN
Clwyd	GBCW
Dyfed	GBDY
Glamorgan	GBGM
Gwent	GBGT
Gwynedd	GBGD
Powys	GBPO

Northern Ireland

Antrim	GBUN
Armagh	GBUR
Down	GBUD
Fermanagh	GBUF
Londonderry	GBUL
Tyrone	GBUT

Channel Islands

Alderney	CIAL
Guernsey	CIGU
Herm	CIHE
Jersey	CIJE
Sark	CISA

Note 5   BTO BBS habitat coding system

BTO HABITAT CODING SCHEME

A WOODLAND

LEVEL 2	LEVEL 3	LEVEL 4*
1 Broadleaved	1 Mixed-aged or semi-natural	1 Dense shrub layer
2 Coniferous	2 Coppice with standards	2 Moderate shrub layer
3 Mixed (10% of each)	3 Coppice without standards	3 Sparse shrub layer
4 Broadleaved water-logged	4 Mature plantation (taller than 10m, with closed canopy)	4 Dense field layer
5 Coniferous water-logged	5 Young plantation (5-10m, open canopy)	5 Moderate field layer
6 Mixed water-logged	6 Parkland (scattered trees and grassy areas)	6 Sparse field layer
	7 High-medium disturbance from people	7 Grazed (moderate to heavy)
	8 Low disturbance	8 Lightly grazed
		9 Dead wood present
		10 Dead wood absent

B SCRUBLAND (or young woodland <5m tall)

LEVEL 2	LEVEL 3	LEVEL 4
1 Regenerating natural or semi-natural woodland	1 Broadleaved	1 Predominantly tall (3-5m)
2 Downland (chalk)	2 Coniferous	2 Predominantly low (1-3m)
3 Heath scrub	3 Mixed (10% of each)	3 Dense shrub layer
4 Young coppice	4 Broadleaved swamp scrub	4 Moderate shrub layer
5 New plantation	5 Coniferous swamp scrub	5 Sparse shrub layer
6 Clear-felled woodland, with or without new saplings	6 Mixed swamp scrub	6 Extensive bracken
7 Other	7 High-medium disturbance from people	7 Dense field layer
	8 Low disturbance	8 Moderate field layer
		9 Sparse field layer
		10 Grazed (moderate to heavy)

C SEMI-NATURAL GRASSLAND/MARSH

LEVEL 2	LEVEL 3	LEVEL 4
1 Chalk downland	1 Hedgerow with trees	1 Ungrazed
2 Grass moor (unenclosed)	2 Hedgerow without trees	2 Cattle
3 Grass moor mixed with heather (unenclosed)	3 Tree-line without hedge	3 Sheep
4 Machair	4 Other field boundary (wall, ditch, etc.)	4 Horses
5 Other dry grassland	5 Isolated group of 1-10 trees	5 Rabbits
6 Water-meadow/grazing marsh	6 No field boundary	6 Deer
7 Reed swamp	7 Montane	7 Other grazers
8 Other open marsh	8 High-medium disturbance from people	8 Extensive bracken
9 Saltmarsh	9 Low disturbance	9 Hay

D HEATHLAND AND BOGS

LEVEL 2	LEVEL 3	LEVEL 4
1 Dry heath	1 Montane	1 Ungrazed
2 Wet heath	2 Raised bog	2 Cattle
3 Mixed heath	3 Valley/basin bog	3 Sheep
4 Bog	4 Blanket bog	4 Horses
5 Breckland	5 Heath mixed with rough grass	5 Rabbits
6 Drained bog	6 Heath without grass	6 Deer
	7 Heath with extensive bracken	7 Other grazers
	8 Undetermined bog	8 Ploughed
	9 Isolated group of 1-10 trees	9 Burned
	10 High-medium disturbance from people	10 Planted with saplings less than 0.5m tall
	11 Low disturbance	

E FARMLAND

LEVEL 2	LEVEL 3	LEVEL 4
1 Improved grassland	1 Hedgerow with trees	1 Ungrazed
2 Unimproved grassland	2 Hedgerow without trees	2 Cattle
3 Mixed grass/tilled land	3 Tree-line without hedge	3 Sheep
4 Tilled land	4 Other field boundary (wall, ditch, etc.)	4 Horses
5 Orchard	5 Isolated group of trees	5 Other stock
6 Other farming	6 Farmyard (active)	6 Bare earth/plough
		7 Autumn cereal
		8 Spring cereal
		9 Root crops (specify)
		10 Other crops (specify)
		11 Oil seed rape
		12 Other brassicas (specify)
		13 Stubble (clean)
		14 Stubble (weedy)
		15 Unsown/Fallow

F HUMAN SITES

LEVEL 2	LEVEL 3	LEVEL 4
1 Urban	1 Building	1 Industrial
2 Suburban	2 Gardens	2 Residential
3 Rural	3 Municipal parks/mown grass/golf courses/recreational areas	3 Well-wooded
	4 Sewage works "urban"	4 Not well-wooded
	5 Near road (within 50m)	5 Area of large gardens
	6 Near active railway line (within 50m)	6 Area of medium gardens
	7 Other	7 Area of small gardens
	8 Rubbish tip	8 Many shrubs
		9 Few shrubs
		10 Disused

G WATER BODIES (freshwater)

LEVEL 2	LEVEL 3	LEVEL 4
1 Pond (less than 50m²)	1 Undisturbed/disused	1 Eutrophic (green water)
2 Small water-body (50-450m²)	2 Water sports (sailing etc)	2 Oligotrophic (clear water, few weeds)
3 Lake/unlined reservoir	3 Angling (coarse or game)	3 Dystrophic (black water)
4 Lined reservoir	4 Coarse angling	4 Marl (clear water, large water-weeds)
5 Gravel pit, sand pit, etc	5 Game fishing	5 Slow-medium running
6 Stream (less than 3m wide)	6 Industrial activity	6 Fast-running
7 River (more than 3m wide)	7 Sewage processing 'rural'	7 Dredged
8 Ditch with water (less than 2m wide)	8 Other disturbance	8 Undredged
9 Small canal (2-5m wide)	9 Small island	9 Banks cleared
10 Large canal (more than 5m wide)		10 Banks vegetated

H COASTAL

LEVEL 2	LEVEL 3	LEVEL 4
1 Marine - open shore	1 Mud or silt	1 Cliff vertical/steeply sloping
2 Marine shore - inlet/cove/loch	2 Sand	2 Dune
3 Estuarine	3 Shingle	3 Flat/gentlysloping
4 Brackish lagoon	4 Rocky	4 Small island
5 Open sea	5 Fully vegetated	5 Spit
	6 Sparse/medium vegetation	6 Dune slack
	7 Inter-tidal	7 Sloping ground
	8 Below low-water mark	8 Undisturbed
		9 Disturbed
		7 Sloping ground
		8 Undisturbed
		9 Disturbed

I INLAND ROCK

LEVEL 2	LEVEL 3	LEVEL 4
1 Cliff	1 Active	1 Bare rock
2 Scree/boulder slope	2 Disused	2 Low vegetation present (mosses, liverworts, etc)
3 Limestone pavement	3 Montane	3 Grasses present
4 Other rock outcrop	4 Non-montane	4 Scrub present
5 Quarry	5 High disturbance from climbers/walkers etc.	
6 Mine/spoil/slag heap	6 Medium disturbance	
7 Cave	7 Low disturbance	

J MISCELLANEOUS

* Shrub layer comprises woody plants less than 5m tall. Field layer comprises herbaceous, non-woody plants.

Note 6 **BTO weather codes for Breeding Birds Survey**

Cloud cover:	0–33% = 1	Rain:	None = 1
	33–66% = 2		Drizzle = 2
	66–100% = 3		Showers = 3
Wind:	Calm = 1	Visibility	Good = 1
	Light = 2		Moderate = 2
	Breezy = 3		Poor = 3

Note 7 **BTO distance categories for Breeding Birds Survey**

Distance from transect line	Distance category
0–25 m	1
25–100 m	2
>100 m (whether in km square or not)	3
Birds in flight only (at any distance)	F

1.16 Core measurement: vertebrates – bats (BA Protocol)

Bats are recorded along a pre-defined transect four times per year, and species and behaviour codes are marked on to maps. Habitat maps are also prepared each year.

Variable	Units	Precision of recording
Map-based recording		
Year of recording		
Habitat	habitat codes[1]	
Visit code	numeric code (1–4)	
Start and end positions of each half of transect	character code (A–D)	
Species	character code[2]	
Behaviour code	character code[3]	
Form-based recording for each half of the transect		
Visit (Sampling) date		
Start time	GMT 24-clock	1 min
Finish time	GMT 24-h clock	1 min
Start position	character code (A–D)	
Start temperature	°C	0.1
Finish temperature	°C	0.1
Cloud cover at start	eighths scale	
Cloud cover at finish	eighths scale	

Recording forms
A recording form is available from the CCU.

Notes
1. Habitat codes (p183).
2. To record the presence of a bat, mark O on the map. If the species is identified, add one of the following species codes to the left of the O:

Noctule	Nn
Serotine	Es
Myotid species	M
Daubenton's	Md
Natterer's	Mn
Pipistrelle	Pp
Long-eared species	P

3. Behaviour codes should be added to the right of the O on the map, and are as follows:

Bat seen but not heard	s
Bat heard echo-locating	h
Feeding buzz heard	f

Note 1   **Habitat codes for bat monitoring**

A. Linear features
1. **Hedgerows**: <4 m high and <5 m wide. All hedgerows are classed as continuous if the gaps are <5 m wide. Mark all larger gaps with a cross-line through the hedgerow and classify each gap as 5–10 m, 11–15 m, 16–20 m, 21–25 m, 26–30 m, 31–35 m, >35 m
2. **Treelines**: a line of single trees (minimum of 3) >4 m high and <2 crown widths apart; continuous and close-knit canopies
3. **Treelines**: as in 2. but discontinuous and spread out
4. **Stone walls**
5. **Footpaths**: small paths, usually only wide enough for one or two people
6. **Tracks/bridleways**: more substantial than above, with an earth or hardcore base but not tarmac
7. **Roads**: tarmac or similar base
8. **Ditches**: usually small, perhaps temporary, watercourses; see 9–12 below
9. **Streams**: flowing water, with no evidence of canalisation, and usually perennial flowing water. Ditches are more likely to dry up and the water flow is more likely to be interrupted
10. **Fast-flowing, rough rivers**
11. **Slow-flowing, smooth rivers**
12. **Canals**: man-made channels

Please note that it is possible to find one or more linear features in association, eg a footpath beside a stone wall, or a tree line in a hedgerow. In such circumstances, both features should be marked on the map.

B. Other habitat features
13. **Semi-natural broadleaved woodland**: predominantly of broadleaved trees >5 m high with a semi-natural appearance
14. **Broadleaved plantations (including orchards)**: tree species not native to the site, even-aged and >5 m
15. **Semi-natural coniferous woodland**: predominantly coniferous trees of any height with semi-natural appearance (confined to Scots pine, juniper and yew)
16. **Coniferous plantations**: predominantly coniferous trees which have been planted and are >5 m
17. **Semi-natural mixed woodland**: at least 25% broadleaved or 25% coniferous trees, of natural appearance with trees >5 m
18. **Mixed plantation**: at least 25% broadleaved or 25% coniferous trees, planted and >5 m
19. **Young plantation**: trees <5 m high, either broadleaved or coniferous, which have been planted
20. **Recently felled woodland**: areas for which there is evidence that woodland has been felled within the past year
21. **Parkland**: areas where tree cover is <30%, the majority of the trees are 30–70 m apart and a minimum of ten trees
22. **Tall scrub**: 3–5 m high

23. **Low scrub**: <3 m high including bracken
24. **Beach**: includes sand dunes, sand/mudflats, shingle or boulder beaches
25. **Lowland heaths**: lowland areas with >25% dwarf shrubs
26. **Heather moorland**: as above but for upland sites
27. **Bog**: areas of peat with vegetation dominated by heather and/or cottongrass
28. **Wet ground**: areas of wet ground found in association with other habitats, eg wet areas in a grass field or flushes in upland areas
29. **Ponds**: up to 0.5 ha
30. **Lakes**: more than 0.5 h
31. **Standing man-made water**: artificially created reservoirs and impoundments, including mill ponds
33. **Upland unimproved grassland**: will include some areas used for rough grazing and poor-quality grassland such as purple moor-grass, unimproved by fertilizers, herbicides or drainage
34. **Lowland unimproved grassland**: may be regularly grazed or mown, but may be totally neglected. Unimproved by fertilizers herbicides or drainage. Includes herb-rich grasslands on limestone, chalk, cliff-tops, etc
35. **Semi-improved grassland**: slightly modified by fertilizer or herbicide application, or by heavy grazing pressure and/or drainage
36. **Improved grassland**: grassland which has had regular treatments of fertilizer and/or herbicide but not including leys (see 40)
37. **Arable**: all classes of arable land, including grassland leys and horticulture. A grassland ley is defined as short-term grassland, and will usually have been re-seeded less than five years previously. It is characterised by evidence of ploughing, bare soil between grass plants, a scarcity of broadleaved plants, and is usually dominated by a single grass species, often rye-grass. There are usually less than 5–10 species per square metre. Category 36 consists of longer-term grassland with a higher density of grass and broadleaved species, usually in enclosed land
38. **Amenity grassland**: includes well-maintained non-agricultural grass
39. **Rock surfaces**
40. **Quarries and open-cast mines**: any excavation (gravel or chalk pits, etc)
41. **Bare soil on unvegetated ground** not falling into 39 or 40.
42. **Built land**: any urban areas including gardens and transport corridors
43. **Others**: please specify

1.17 Core measurement: vertebrates – rabbits and deer (BU Protocol)

Rabbits and deer are recorded *via* dropping counts in sections along pre-defined transects twice per year, in late March and late September. Each transect is regarded as a single sampling location (code 01, 02, etc).

Variable	Units	Precision of recording
Date droppings cleared		
Date droppings recorded (Sampling date)		
Transect section	character code (Tn)[1]	
Number of rabbit droppings	count	1
Number of deer droppings	count	1
Transect section length	m	0.1
Habitat description		

Recording forms
Two recording forms are available from the CCU, one for recording droppings of rabbits and deer, and one to describe lengths and habitat types for each section. The second of these is to be completed on the first year of survey, and again if any habitat change occurs, or transects sections have to be altered. The number of transect sections used depends on the habitat pattern at each site.

Note
1. Transect sections should be numbered sequentially across both transects so that they are unique within each ECN site.

1.18 Core measurement: vertebrates – frog spawn (BF Protocol)

Frog spawn monitoring is conducted annually at selected ponds and ditches, taking daily measurements of spawn and weekly measurements of pond environment. A single sample of pond water, taken at the time of first spawning, is analysed annually for a set of chemical determinands (see below). The pond is checked weekly from the 1 January for signs of frogs congregating and calling. Daily records are made of spawn development from the time the frogs are first seen congregating until the first eggs have hatched. Weekly records of pond pH and temperature are made between the date of first spawning and the date on which the newly metamorphosed frogs are first seen leaving the pond.

Variable	Units	Precision of recording
Pond ID	as location code (01,02, etc)	
Date frogs first seen congregating		
Date of first spawning		
Date of first hatching		
Date newly metamorphosed frogs first seen leaving		

Pond sample
Sampling date

pH	pH scale	0.1
Conductivity	µS cm^{-1}	0.1
Alkalinity	mg l^{-1}	3 significant figures
Na$^+$	mg l^{-1}	3 significant figures
K$^+$	mg l^{-1}	3 significant figures
Ca^{2+}	mg l^{-1}	3 significant figures
Mg^{2+}	mg l^{-1}	3 significant figures
Fe^{2+}	mg l^{-1}	3 significant figures
Al^{3+}	mg l^{-1}	3 significant figures
PO$_4$$^{3-}$–P	mg l^{-1}	3 significant figures
NH$_4$$^+$–N	mg l^{-1}	3 significant figures
Cl$^-$	mg l^{-1}	3 significant figures
NO$_3$$^-$–N	mg l^{-1}	3 significant figures
SO$_4$$^{2-}$–S	mg l^{-1}	3 significant figures
Dissolved Organic Carbon	mg l^{-1}	3 significant figures

Pond monitoring (weekly)
Recording (Sampling) date
Date Min/Max thermometers set

Depth at centre of spawning area	cm	0.1
Minimum temperature	°C	1
Maximum temperature	°C	1
pH (at Position 1)	pH scale	0.1
pH (at Position 2)	pH scale	0.1
pH (at Position 3)	pH scale	0.1

Spawn monitoring (daily)
Recording (Sampling) date
Date of last recording

Number of new spawn masses	count	1
Total surface area of pond covered by spawn	m^2	0.1
Percentage dead or diseased eggs	%	1

Recording forms
A standard field recording form is available from the CCU. An example is
provided in Appendix II.

2. Guidelines and formats for transfer of data to the ECN CCU

This Section describes the general procedures used in transferring machine-readable data to the CCU for validation and input to the central database. Detailed documentation describing the specific transfer formats for each core measurement are not provided here, but can be obtained from the CCU.

ECN has adopted the term 'core measurement' to mean an aspect of the environment on which a set of measurements will be made, eg vegetation, soils, invertebrates. Each ECN site has been assigned a 'Site Identification Code', and each ECN core measurement (or subcategory of a core measurement) a 'Measurement Code' which are referenced in the database and in datasets transferred to the CCU. These identifiers are listed in Appendix I.

2.1 Data media

Data are sent over the Internet by electronic mail (e-mail), but may also be sent by diskette where no Internet connection is available.

2.1.1 Transfer by electronic mail

Data should be sent to the e-mail address: ECN@ITE.AC.UK

Each e-mail dataset (message) should normally contain data for only one ECN core measurement and one ECN site. The dataset should begin with header information, as described in Section 2.2.1 below and terminate with END on a separate line. The subject field of each message is used to describe uniquely its contents, including in particular the site and the measurement code to which the data relate. For example: 'Wytham MA data, Jan–Mar 1994'.

The first e-mail message of a set should be an **information message** containing:
> Site name, Site ID code
> Name of person sending the e-mail
> Date
> A list of the data messages to follow (as described by the subject field of each)
> Any additional text information necessary to qualify each dataset (see Section 2.4)

2.1.2 Disk transfer

Disks should be sent to ECN Database Manager, ITE Merlewood Research Station, Grange-over-Sands, Cumbria LA11 6JU, UK.

IBM-format 3.5 inch diskettes should be used, labelled with:
> Site name, Site ID code

Name of the person sending the disk
Date sent
Core measurement codes contained on the disk

Each data file should normally contain data for only one ECN core measurement for one ECN site, and begin with header information, as described in Section 2.2.1 below. Data file names should reflect the data they contain, eg MA96-1.DAT for automatic weather station data for the first quarter of 1996.

The disk should also include a file called '**inform.txt**', which contains the name of each data file and its respective ECN core measurement code (see Appendix I) and dates, followed by any text necessary to qualify the data in that data file (see Section 2.4).

2.2 Data file format

The wide variety of software systems in use by ECN organisations has made it difficult to standardise on specific software for data transfer. The most straightforward solution has been to standardise on ASCII text files in comma-separated format, which most software products are able to support. Data entry forms with automatic validation checks are currently being developed for use by those organisations without their own databases and associated data entry software.

Each data record is uniquely referenced in time and space within the ECN database. The logical structure of datasets is different for each ECN core measurement, and this affects the format for data transfer. For most transfer datasets, each individual data record will include a date reference – the Sampling Date – as one of its key fields. A spatial reference to site and internal location code must also be provided; this may either be part of the header information or one of the fields in the data record, depending on whether the whole dataset refers to a single location (eg hourly data from one AWS) or more than one (eg different stream water samples). These details are provided in the data transfer documentation for each core measurement, available from the CCU.

2.2.1 Data file header

Each data file should normally begin with four lines of header information, as follows:

Measurement code	(eg MM)
Site name	(eg North Wyke)
Site ID code	(eg 05)
Internal location code(s)	(eg 01) and/or additional specified information

The 'Location Code' is a numeric code which may be required to identify a particular instrument, TSS or location where an ECN site has more than one. Where only one exists (eg most sites have only one AWS), it should be assigned the code 01. Where the dataset itself includes information from a number of locations (eg stream sites), then the location codes should be listed on the fourth header line, separated by commas, eg 01,02,03 for three stream sites. Some core measurements may require additional information to be supplied on this line. Data transfer documents for individual core measurements will indicate in more detail what is expected here.

2.2.2 Data record format

The data transfer document for each core measurement gives details of the required standard format for data records. A general principle is that data should be in 'free-format' with comma-separated fields in the order specified in the transfer document. Quality codes describing any problems relating to a given sampling date may be attached to each data record. Section 2.4 describes the procedures for supplying quality codes and text in more detail.

2.2.3 Zero values

Data values of zero are equally as important as non-zero values, and should be transmitted to the ECN database in the same way. It is also important that the distinction between zero values and missing data (see Section 2.3) is understood. For example, if there is no water in a precipitation collector at a given sampling date, the volume must be recorded as *zero*, not as missing.

2.2.4 Dataset continuity

A dataset sent to the CCU for a given core measurement should continue logically from the final record of the last dataset sent, so that there are no gaps or overlaps. For example, if the final record of an automatic weather station dataset is for 1100 on 1 February, then the first record of the next dataset sent should be for 1200 on 1 February. If for some reason there are breaks in the data, these should be explained in the accompanying 'inform.txt' file or e-mail message (see Section 2.4).

2.3 Missing data and non-standard sampling dates

2.3.1 Missing data fields

Data values should be recorded as 'missing' where a reading or sample was not able to be taken, eg if an instrument has broken down. Missing data values should be recorded as 'null' fields by simply including the separating comma in the data record where otherwise the data value would be. It is most important that these separating commas appear in the correct position in the data record; otherwise it will not be clear which data field refers to which variable. Information about the reasons for missing data should be given either through the quality codes attached to the data records or, if no code is suitable, in the 'inform.txt' file or first e-mail message, accompanied by dates or date ranges (see Section 2.4 below).

2.3.2 Missing data records

It is important that ECN samples or recordings are made on the standard sampling due dates and according to the sampling periods specified in the Protocols. However, there may be occasions where sampling or recording is not possible, and all data fields are effectively missing. As a general rule, data records for **all dates on which sampling/recording is due**, even if missing, should be included in the dataset, with null data fields, and appropriate quality codes and/or text supplied (see Section 4 below). Exceptions to this rule, where records may be omitted, are:

- if there are runs of more than three successive missing sampling due dates, eg through instrumentation failure. However, information on why dates are missing should be included in the first e-mail message or 'inform.txt' file (see Section 2.4).
- if another date is effectively substituted for the sampling due date – for example, if a surveyor is prevented from making a recording because of bad weather, but successfully attempts to make the recording the following day. In this case, *only* the data record for the substitute date should be included, *omitting* that for the sampling due date. However, the quality code 222 'non-standard sampling date' should be attached to the substitute data record, and reasons given in the first e-mail message or 'inform.txt' file (see Section 2.4).
- laboratory analysis results. Only those data records for which a field sample has been provided should be included. (Information about missing samples will be provided through their associated field sampling datasets.)

2.4 Data quality

2.4.1 Supplying information about quality

It is important that ECN data are accompanied by information about their quality. Standard operating procedures and target quality criteria are specified in the measurement protocols and specifications are stored in the ECN database. Wherever a value has been recorded using a different procedure or specification, or where other factors beyond direct control affect recording or sampling, this information needs to accompany the data.

Where appropriate, pre-defined 'quality codes' which describe common sampling problems can be included on the end of data **records,** ie they apply to a given sampling date at a single location. A list of quality codes has been drawn up (see Appendix III) which uses a three-figure numeric sequence; codes are added to the list if problems continue to recur. Examples are:

129   Condensation in diffusion tube
135   Pitfall trap flooded
101   No sample/recording taken – equipment out of action/unable to visit equipment
222   Non-standard sampling date

999   Free-text information is associated with this data record
000   No problems with sampling/recording – no quality codes or text apply

If a problem occurs with a particular measurement for which there is no existing code, then free text comments should be recorded on the field form and reported in the first e-mail message or 'inform.txt' file when transferring data to the CCU; quality code 999 should also be attached to the data record to indicate that text information has been supplied. New codes are allocated in consultation with the CCU if the reported condition recurs regularly, and a new list issued. Those quality codes particularly appropriate to a given core measurement are listed on its associated field form and in its accompanying data transfer document. Any number of codes can be appended to a single data record. Quality code 000 should be used where quality issues have been considered but no quality information applies (either as codes or text). This provides positive differentiation from the situation where information about quality has not been considered at all (assumed if no code at all appears at the end of a data record).

Where data fields relate to time periods, eg total rainfall, the length of the time period needs to accompany the data. Most data transfer formats relating to a sampling period incorporate date/time ranges within each sampling data record. However, there may be occasions when selected variables run over longer time periods than others, eg where a recorder forgets to measure the contents of the raingauge and therefore the following day's reading relates to two days' accumulation. It is particularly important in these cases that appropriate quality codes and information are supplied indicating non-standard time periods due to omissions, instrumentation breakdown or inoperation between sampling dates.

2.4.2 Data checking and validation

It is most important that individual sites perform basic typing and formatting checks on the dataset before it is sent to the CCU. Examples are:
- data reported in the specified format,
- data fields in the specified order,
- missing fields correctly indicated, and
- correct quality codes used where necessary.

Validation for categorical codes and numeric data ranges will be made as data are entered into the ECN database. However, values which fall outside the ranges will only be discarded if there is a clear explanation, such as instrumentation error, and corrections made where possible. If the reason is unclear, the values will be stored, but qualified in the meta-database. For these reasons, recorders are asked not to alter data unless there is a clear explanation for error. Any alterations made on this basis (apart of course from typing and formatting errors), or any suspicions about the validity of the data should be detailed in the first e-mail message or 'inform.txt' file.

2.5 Frequency of data transfer to the CCU

How often data are transferred to the CCU and the time-lag between data capture and data transfer will depend to some extent on the core measurement Protocol. It is necessary to strike a balance between keeping the database as up-to-date as possible, validating data as quickly as possible, and on the other hand maintaining a realistic level of workload for all those involved. In addition, the CCU has a responsibility to produce summary statistics and carry out analyses to meet agreed schedules. It is suggested that quarterly transfers of data with up to a one month time-lag would be suitable for those core measurements which are monitored throughout the year. Information about frequency and time-lag for data transfer is included in the respective transfer documents.

2.6 Data receipt and data backup

Data received by the CCU are acknowledged when safely transferred into the database. Disks are returned to sites in batches. Suitable arrangements have been made for backing up the ECN database itself. However, it will be particularly important to ensure that data are secure during the time between collection and entry into the database. For this reason, sites should ensure that they keep copies of the data at least until safe storage is acknowledged. Sites are asked also to keep hard copy recording forms for as long as is practicable, as an ultimate reference for any information which may have been missed when coding and entering data.

A.M.J. Lane
ECN Database Manager

References Avery, B.W. 1980. *Soil classification for England and Wales.* (Technical Monograph no. 14.) Harpenden: Soils Survey.

Barr, C.J., Bunce, R.G.H., Clarke, R.T., Fuller, R.M., Furse, M.T., Gillespie, M.K., Groom, G.B., Hallam, C.J., Hornung, M., Howard, D.C. & Ness, M.J. 1993. *Countryside Survey 1990: main report.* (Countryside Survey 1990 vol.2.) London: Department of the Environment.

Heslop, I.R.P. 1964. *Revised indexed checklist of the British Lepidoptera library edition.* (Reprinted from the *Entomologist's Gazette*, 1959–63.)

Hodgson, J.M. 1974. *Soil Survey field handbook.* Harpenden: Soil Survey.

Malloch, A.J.C. 1988. *VESPAN II: A computer package to handle and analyse multivariate species data and handle and display species distribution data.* Lancaster: Institute of Environmental and Biological Sciences, University of Lancaster.

Meteorological Office. 1982. *Observer's handbook.* London: HMSO.

Rodwell, J.S. 1991 et seq. *British plant communities.* Cambridge: Cambridge University Press

APPENDIX I   ECN terrestrial site ID codes and core measurement codes

SITE IDENTIFICATION CODES – terrestrial sites (1996)

01	DRAYTON	07	SOURHOPE
02	GLENSAUGH	08	WYTHAM
03	HILLSBOROUGH	09	ALICE HOLT
04	MOOR HOUSE/UPPER TEESDALE	10	PORTON DOWN
05	NORTH WYKE	11	Y WYDDFA/SNOWDON
06	ROTHAMSTED		

CORE MEASUREMENT CODES

MA   Meteorology: AWS
MM   Meteorology: manual

AC   Atmospheric chemistry

PC   Precipitation chemistry

WD   Surface water: discharge
WC   Surface water: chemistry and quality

SB   Soil survey and classification: baseline data
SC   Soil characterisation and change – coarse-grain monitoring (every 20 years)
SF   Soil characterisation and change – fine-grain monitoring (every five years)

SS   Soil solution chemistry

VB   Vegetation – baseline survey
VC   Vegetation – coarse-grain sampling
VF   Vegetation – fine-grain sampling
VW   Vegetation – woodlands
VH   Vegetation – hedgerows and linear features
VP   Vegetation – permanent pasture
VA   Vegetation – cereals

IT   Invertebrates: tipulids
IM   Invertebrates: moths
IB   Invertebrates: butterflies
IS   Invertebrates: spittle bugs
IG   Invertebrates: ground predators

BC   Vertebrates, common birds
BB   Vertebrates, breeding birds
BM   Vertebrates, moorland birds
BA   Vertebrates, mammals: bats
BU   Vertebrates, mammals: rabbits and deer
BF   Vertebrates, amphibians: frog spawn

APPENDIX II Example recording forms for ECN Protocols

Some example recording forms (*) are provided in this Appendix. The full set, listed below, can be obtained from the CCU.

Water chemistry (WC) – 1 recording form (*)
Precipitation chemistry (PC) – 1 recording form
Soil solution chemistry (SS) – 1 recording form
Atmospheric chemistry (AC) – 1 recording form (*)
Vegetation (V) – 15 recording forms (* VB only)
Tipulids (ST) – 3 recording forms
Butterflies (IB) – 1 recording form (from Butterfly Monitoring Scheme)
Spittle bugs (IS) – 3 recording forms
Ground predators (IG) – 5 recording forms (* 2)
Common Birds Census (BC) – 2 summary sheets (from British Trust for Ornithology)
Breeding Bird Survey (BB) – 3 recording forms (from British Trust for Ornithology)
Moorland birds (BM) – 1 recording form
Bats (BA) – 1 recording form
Rabbits and deer (BU) – 2 recording forms
Frog spawn (BF) – 1 recording form (*)

Field recording form sample

FIELD RECORDING FORM SAMPLE:

ENVIRONMENTAL CHANGE NETWORK: SURFACE WATER CHEMISTRY (WC)
(All dates to be given in dd-mon-yyyy format. All times to be given in GMT hh:mm format)

Site Name: ....Site ID:..............

Collection Date:..

Weather...

Collected By..

Sample Code			Time (hh:mm)	River Stage(mm)	Quality Codes	Quality Comments
Mmt Code	Site ID	Location Code				
WC		01				
WC		02				
WC		03				
WC		04				

Field Quality Codes Most Relevant to this Measurement

101	No sample - equipment out of action / unable to visit equipment
102	Sample lost or inadvertently discarded
103	Partial loss of sample
106	Snow during sampling period
113	Sample discoloured
114	Bonfire in vicinity during sampling period
115	Heather burning in vicinity during sampling period
116	Forest fire in vicinity during sampling period
117	Straw burning in vicinity during sampling period
118	Crop spraying in vicinity during sampling period
119	Construction work in vicinity during sampling period
120	Liming in vicinity during sampling period
126	River frozen - no sample
127	River dry - no sample

ECN/FRM/WC/27.8.96

Field recording form sample

FIELD RECORDING FORM SAMPLE

ENVIRONMENTAL CHANGE NETWORK: ATMOSPHERIC CHEMISTRY
(All dates to be given in dd-mon-yyyy formet. All times to be given in GMT hh:mm format)

Site Name:..Site ID:................ Location Code:.....................

[Date Tubes originally set out:..Time set out:............................]

Date Tubes Collected:...Collection Time:.....................................

Weather...

Tubes Collected by:...

Sample Code			Quality Codes	Quality Comments
Mmt Code	Site ID	Tube Code		
AC		E1		
AC		E2		
AC		E3		
AC		B1		
AC		B2		
AC		B3		

Tube Codes

E   Experimental Tube
B   Blank Tube

Field Quality Codes Most Relevant to this Measurement

101   No sample - equipment out of action / unable to visit equipment
102   Sample (tube) lost or inadvertently discarded
114   Bonfire in vicinity during sampling period
115   Heather burning in vicinity during sampling period
116   Forest fire in vicinity during sampling period
117   Straw burning in vicinity during sampling period
128   Fauna in tube
129   Condensation in tube

Date **New** tubes set out:..Time **New** tubes set out:................

ECN/FRM/AC/27.8.96

197

Vegetation baseline recording form sample

ENVIRONMENTAL CHANGE NETWORK: BASELINE VEGETATION RECORDING (VB) - Species Data

Site Name:.. Date (dd-mon-yyyy):...

Plot Position ID:.. Name of Surveyor:..

Standard baseline 2m x 2m plot (S) or Woodland 10m x 10m plot (W) ?........................

2m

▲ 2m
N

(or 10m x 10m for woodland recording)

Species	Code	Pres.	Species	Code	Pres.	Species	Code	Pres.	Species	Code	Pres.	Species	Code	Pres.
Acer pseud (c)	103		Clad impex	2366		Gera rober	630		Oxal aceto	932		Rume obtus	1147	
Achi mille	104		Clad pyxid	2379		Geum urban	634		Pedi sylva	947		Sagi sp.	4291	
Achi ptarm	105		Clad uncia	2391		Glec heder	637		Pell sp.	2917		Samb nigra (s)	1187	
Agro canin	120		Cono majus	431		Hede helix (g)	652		Phle prate	960		Scir cespi	1210	
Agro capil	123		Conv arven	433		Hera sphon	661		Pice sitch (c)	3122		Sene jacob	1239	
Agro stolo	122		Cory avell (s)	441		Hier pilos	3635		Ping vulga	970		Sene vulga	1243	
Alli petio	144		Crat monog (s)	445		Hier sp	675		Plag undul	1868		Sile dioic	1254	
Alop genic	156		Crep sp.	2692		Holc lanat	680		Plan lance	973		Sonc asper	1272	
Alop prate	158		Cyno crist	460		Holc molli	681		Plan major	974		Sonc olera	1273	
Ange sylve	167		Dact glome	465		Hord vulga	2854		Pleu schre	1872		Sorb aucup (c)	1275	
Anth odora	171		Dact macul	468		Hyac nonsc	516		Poa annua	981		Spha capil	1960	
Anth sylve	173		Dant decum	1249		Hylo splen	1761		Poa prate	988		Spha cuspi	1963	
Arrh elati	197		Desc cespi	477		Hype pulch	702		Poa trivi	990		Spha palus	1971	
Arum macul	201		Desc flexu	478		Hypn cupre	1766		Poly vulga	995		Spha papil	1972	
Athy filix	215		Dicr heter	1615		Hypo sp.	4377		Poly avicu	999		Spha recur	1976	
Aven fatua	222		Dicr scopa	1638		Junc acuti	719		Poly persi	1008		Stac sylva	1293	
Bell peren	230		Digi purpu	482		Junc artic	722		Poly commu	1891		Stel alsin	1295	
Betu pendu (c)	237		Dros rotun	494		Junc bulbo	726		Poly junip	1894		Stel holos	1297	
Betu pubes (c)	236		Dryo dilat	499		Junc congl	729		Pote anser	1043		Stel media	1298	
Blec spica	242		Dryo filix	500		Junc effus	730		Pote erect	1046		Succ prate	1305	
Brac sylva	247		Elym repen	118		Junc squar	736		Pote repta	1050		Tara agg.	2982	
Brac sp.	3046		Empe nigru	515		Lami album	749		Prim vulga	1058		Thui tamar	2003	
Brom molli	258		Epil angus	391		Lami purpu	753		Prun spino (s)	1065		Thym praec	1333	
Brom steri	262		Epil hirsu	521		Laps commu	754		Prun vulga	1059		Tori japon	1337	
Call vulga	278		Epil monta	522		Lath prate	758		Pseu purum	1914		Trif dubiu	1343	
Camp rotun	288		Epil palus	525		Leo sp.	4297		Pter aquil	1066		Trif prate	1349	
Caps bursa	290		Equi arven	532		Loli pulch	795		Quer petra (c)	1077		Trif repen	1350	
Card prate	295		Eric ciner	541		Loli peren	796		Quer robur (c)	1078		Trit aesti	2989	
Care biner	308		Eric tetra	542		Loni peric (g)	798		Raco lanug	1932		Ulex europ (s)	1363	
Care demis	312		Erio angus	546		Loph sp.	2871		Ranu acris	1081		Urti dioic	1368	
Care echin	319		Erio vagin	548		Lotu corni	800		Ranu ficar	1088		Vacc myrti	1375	
Care nigra	333		Euph sp.	568		Luzu campe	807		Ranu flamm	1089		Vero arven	1393	
Care panic	339		Eurh sp.	4319		Luzu multi	809		Ranu repen	1095		Vero chama	1396	
Care pilul	344		Fest ovina	574		Matr matri	839		Rhyt loreu	1939		Vero offic	1401	
Cent nigra	371		Fest rubra	576		Merc peren	864		Rhyt squar	1940		Vero persi	1403	
Cera fonta	384		Fili uimar	583		Mniu hornu	1794		Rosa sp.	1122		Vero serpy	1406	
Chen album	395		Frax excel (c)	589		Mniu undul	1807		Rubu fruti agg	1136		Vici sepiu	1416	
Chry oppos	408		Gali apari	605		Moli caeru	876		Rume a'sa	1139		Viol palus	1427	
Cirs arven	415		Gali palus	609		Myri gale	893		Rume a'la	1140		Viol riv/rei	4376	
Cirs palus	418		Gali saxat	610		Nard stric	900		Rume congl	1142				
Cirs vulga	419		Gera molle	625		Nart ossif	901		Rume crisp	1143				

OTHER SPECIES	CODE	Presence	OTHER SPECIES	CODE	Presence

(200 most common species derived from ITE Merlewood Countryside Survey 1990)

ECN/FRM/VB1/20.5.93

Vegetation baseline recording form sample

ENVIRONMENTAL CHANGE NETWORK: BASELINE VEGETATION RECORDING (VB) - DESCRIPTIVE AND QUALITY INFORMATION

Site Name:... Date (dd-mon-yyyy).................. Name of Surveyor:..........................

Notes: (1) 'S' or 'W' in column 2 below refers to Standard Baseline (2m x 2m) plot or Woodland (10m x 10m) plot. Where both types of plot have been surveyed, 2 entries will be needed in the table.

(2) For columns PATH through to Natural Boundary below, tick as appropriate

2m

▲ N
2m

(or 10m x 10m if a Woodland plot)

Plot Posn. ID	'S' or 'W'	PATH	WALL	DITCH	BANK	FENCE	STREAM	HEDGE	Natural Boundary	If plot obstructed give Percentage Unsurveyed	Quality Codes	Comments

ECN/FRM/VB224.5.93

Quality codes most relevant to vegetation recording

130   Plot/transect/cell not surveyed
136   Cell not surveyed due to obstruction
132   Plot/transect/cell not accurately relocated
131   Trampling in plot/transect/cell
137   Flooding in plot/transect/cell
115   Heather burning in plot/transect/cell
117   Straw burning in plot/transect/cell
116   Forest fire in plot/transect/cell
133   Evidence of disease in plot/transect/cell
134   Significant disturbance in plot/transect/cell
138   Mowing in plot/transect/cell
120   Liming in plot/transect/cell
139   Muck/slurry/slag application in plot/transect/cell
140   Application of chemicals (eg fertilizers/sprays) in plot/transect/cell
141   Grazing by sheep in plot/transect/cell
142   Grazing by cattle in plot/transect/cell
143   Grazing/browsing by deer in plot/transect/cell
144   Grazing – other in plot/transect/cell
145   Woodland management: coppicing in plot/transect/cell
146   Woodland management: thinning in plot/transect/cell
147   Woodland management: clearfelling in plot/transect/cell
148   Woodland management: brashing in plot/transect/cell
149   Wind-throw in plot/transect/cell
150   Animal path present in plot/transect/cell
151   Forest ride present in plot/transect/cell
152   Human path present in plot/transect/cell
153   Path present (unspecified) in plot/transect/cell
200   Adverse weather conditions affected survey
201   Biting insects affected survey
202   Failing light affected survey

Note
Codes 115,117,138,120,139 and 140 should be followed up with details
of farming practices as part of the site management information – in
particular, actual chemicals used, how and when they are applied.

Full names and codes of the 200 species used on the vegetation recording forms

(Taken from the ITE Countryside Survey Database as the 200 most common species in GB)

103 Acer pseudoplatanus (c)
104 Achillea millefolium
105 Achillea ptarmica
120 Agrostis canina
123 Agrostis capillaris
122 Agrostis stolonifera
144 Alliaria petiolata
156 Alopecurus geniculatus
158 Alopecurus pratensis
167 Angelica sylvestris
171 Anthoxanthum odoratum
173 Anthriscus sylvestris
197 Arrhenatherum elatius
201 Arum maculatum
215 Athyrium filix-femina
222 Avena fatua
230 Bellis perennis
237 Betula pendula (c)
236 Betula pubescens (c)
242 Blechnum spicant
247 Brachypodium sylvaticum
3046 Brachythecium sp.
258 Bromus mollis (syn)
262 Bromus sterilis
278 Calluna vulgaris
288 Campanula rotundifolia
290 Capsella bursa-pastoris
295 Cardamine pratensis
308 Carex binervis
312 Carex demissa
319 Carex echinata
333 Carex nigra
339 Carex panicea
344 Carex pilulifera
371 Centaurea nigra
384 Cerastium fontanum triviale
395 Chenopodium album
408 Chrysosplenium oppositifolium
415 Cirsium arvense
418 Cirsium palustre
419 Cirsium vulgare
2366 Cladonia impexa
2379 Cladonia pyxidata
2391 Cladonia uncialis
431 Conopodium majus

433 Convolvulus arvensis
441 Corylus avellana (s)
445 Crataegus monogyna (s)
2692 Crepis sp.
460 Cynosurus cristatus
465 Dactylis glomerata
468 Dactylorhiza maculata maculata
1249 Danthonia decumbens
477 Deschampsia cespitosa cespitosa
478 Deschampsia flexuosa
1615 Dicranella heteromalla
1638 Dicranum scoparium
482 Digitalis purpurea
494 Drosera rotundifolia
499 Dryopteris dilatata
500 Dryopteris filix-mas
118 Elymus repens
515 Empetrum nigrum nigrum
391 Epilobium angustifolium
521 Epilobium hirsutum
522 Epilobium montanum
525 Epilobium palustre
532 Equisetum arvense
541 Erica cinerea
542 Erica tetralix
546 Eriophorum angustifolium
548 Eriophorum vaginatum
568 Euphrasia sp.
4319 Eurhynchium sp.
574 Festuca ovina
576 Festuca rubra
583 Filipendula ulmaria
589 Fraxinus excelsior (c)
605 Galium aparine
609 Galium palustre
610 Galium saxatile
625 Geranium molle
630 Geranium robertianum
634 Geum urbanum
637 Glechoma hederacea
652 Hedera helix (g)
661 Heracleum sphondylium
3635 Hieracium pilosella (pil)
675 Hieracium sp.
680 Holcus lanatus

681 Holcus mollis
2854 Hordeum vulgare
516 Hyacinthoides nonscripta
1761 Hylocomium splendens
702 Hypericum pulchrum
1766 Hypnum cupressiforme
4377 Hypochoeris sp.
719 Juncus acutiflorus
722 Juncus articulatus
726 Juncus bulbosus
729 Juncus conglomeratus
730 Juncus effusus
736 Juncus squarrosus
749 Lamium album
753 Lamium purpureum
754 Lapsana communis
758 Lathyrus pratensis
4297 Leontodon sp.
795 Lolium multiflorum
796 Lolium perenne
798 Lonicera periclymenum (g)
2871 Loph sp.
800 Lotus corniculatus
807 Luzula campestris
809 Luzula multiflora
839 Matricaria matricarioides (syn)
864 Mercurialis perennis
1794 Mnium hornum
1807 Mnium undulatum (syn)
876 Molinia caerulea
893 Myrica gale
900 Nardus stricta
901 Narthecium ossifragum
932 Oxalis acetosella
947 Pedicularis sylvatica
2917 Pellia sp.
960 Phleum pratense
3122 Picea sitchensis (c)
970 Pinguicula vulgaris
1868 Plagiothecium undulatum
973 Plantago lanceolata
974 Plantago major
1872 Pleurozium schreberi
981 Poa annua
988 Poa pratensis
990 Poa trivialis
995 Polygala vulgaris
999 Polygonum aviculare
1008 Polygonum persicaria
1891 Polytrichum commune

1894 Polytrichum juniperinum
1043 Potentilla anserina
1046 Potentilla erecta
1050 Potentilla reptans
1058 Primula vulgaris
1059 Prunella vulgaris
1065 Prunus spinosa (s)
1914 Pseudoscleropodium purum
1066 Pteridium aquilinum
1077 Quercus petraea (c)
1078 Quercus robur (c)
1932 Racomitrium lanuginosum
1081 Ranunculus acris
1088 Ranunculus ficaria
1089 Ranunculus flammula
1095 Ranunculus repens
1939 Rhytidiadelphus loreus
1940 Rhytidiadelphus squarrosus
1122 Rosa sp.
1136 Rubus fruticosus agg.
1139 Rumex acetosa
1140 Rumex acetosella
1142 Rumex conglomeratus
1143 Rumex crispus
1147 Rumex obtusifolius
4291 Sagina sp.
1187 Sambucus nigra (s)
1210 Scirpus cespitosus
1239 Senecio jacobaea
1243 Senecio vulgaris
1254 Silene dioica
1272 Sonchus asper
1273 Sonchus oleraceus
1275 Sorbus aucuparia (c)
1960 Sphagnum capillifolium
1963 Sphagnum cuspidatum
1971 Sphagnum palustre
1972 Sphagnum papillosum
1976 Sphagnum recurvum
1293 Stachys sylvatica
1295 Stellaria alsine
1297 Stellaria holostea
1298 Stellaria media
1305 Succisa pratensis
2982 Taraxacum agg.
2003 Thuidium tamariscinum
1333 Thymus praecox arcticus
1337 Torilis japonica
1343 Trifolium dubium
1349 Trifolium pratense

1350 Trifolium repens
2989 Triticum aestivum
1363 Ulex europaeus (s)
1368 Urtica dioica
1375 Vaccinium myrtillus
1393 Veronica arvensis
1396 Veronica chamaedrys
1401 Veronica officinalis
1403 Veronica persica
1406 Veronica serpyllifolia serpyllifolia
1416 Vicia sepium
1427 Viola palustris
4376 Viola riviniana/reichenb

Ground predator recording form sample

ENVIRONMENTAL CHANGE NETWORK: GROUND PREDATORS (IG) - TRANSECT 01

IG₁

Site Name:...   Site ID:................   Location (Transect Code):.......01.........

Date traps set out:...   Collected by :..

Date traps emptied:...   Identified by :..

Species	Spec. Code	Trap 1	Trap 2	Trap 3	Trap 4	Trap 5	Trap 6	Trap 7	Trap 8	Trap 9	Trap10

Ground predator recording form sample

ENVIRONMENTAL CHANGE NETWORK: GROUND PREDATORS (IG)

PITFALL TRAP QUALITY INFORMATION

Site Name:.. Collected by:...

Date traps emptied (dd-mon-yyyy):..

Trap ID No.	Transect Code	Quality Codes	Quality Comments
1	01		
2	01		
3	01		
4	01		
5	01		
6	01		
7	01		
8	01		
9	01		
10	01		
11	02		
12	02		
13	02		
14	02		
15	02		
16	02		
17	02		
18	02		
19	02		
20	02		
21	03		
22	03		
23	03		
24	03		
25	03		
26	03		
27	03		
28	03		
29	03		
30	03		

ECN/FRM/IG4/6.6.96

Quality codes most relevant for ground predator recording

101 No sample – equipment out of action/unable to visit equipment
102 Sample lost or inadvertently discarded
103 Partial loss of sample (through damage to trap or other reasons)
106 Snow during sampling period
134 Significant disturbance in vicinity of trap
135 Pitfall trap flooded
115 Heather burning affected sample area during sampling period
116 Forest fire affected sample area during sampling period
117 Straw burning affected sample area during sampling period
118 Crop spraying affected sample area during sampling period
119 Construction work affected sample area during sampling period
120 Liming affected sample area during sampling period
121 Change of land use affected sample area during sampling period
160 Pitfall trap damaged
161 Pitfall trap lost
162 No antifreeze left in pitfall trap
163 Trampling by cattle during sampling period
164 Pitfall trap contained enough debris/mud to affect catch
165 Pitfall trap raised above ground level
166 Small amount of mud/debris in pitfall trap may have obscured small ground predators
167 Pitfall trap lid removed during sampling period
168 Trampling by sheep during sampling period
212 Small mammal(s) in trap – species unrecorded
213 Short-tailed field vole(s) in trap
214 Pygmy shrew(s) in trap
215 Frog(s) in trap
216 Common shrew(s) in trap
217 Common lizard(s) in trap

Frog spawn recording form sample

BF

FIELD RECORDING FORM SAMPLE:

ENVIRONMENTAL CHANGE NETWORK: FROG SPAWN (BF)

Site Name:................................ Site ID:................ Date of Recording:................ Recorded by:................

Pond ID	Time of Sampling (hh:mm)	Number of new Spawn Masses	Total Surface Area of Spawn (m)	% Dead / Diseased Eggs	Depth (cm)	Min Temp (oC)	Max Temp (oC)	pH1	pH2	pH3	Sample Taken (Y or N)	Quality Codes/Comments
01												
02												
03												
04												
05												

Field Quality Codes Most Relevant to this Measurement:

101	No sample/reading taken - unable to visit sampling site
102	Sample lost or inadvertantly discarded
103	Partial loss of sample
106	Snow during sampling period
111	Unidentified debris in sample
113	Sample discoloured
170	Pond frozen
171	Spawn possibly killed by frost
172	Number of spawn masses to large to count
173	Number of spawn masses approximate
174	Pond dried up
175	No frogs seen prior to spawning
176	Pond covered with snow
177	No newly metamorphosed frogs seen leaving pond
178	Spawn stranded and dried out due to falling water levels
179	Pond drying up - only small surface puddles left
200	Adverse weather conditions affected recording

APPENDIX III   ECN quality codes for survey

(Current at 10th June 1996)

100   No information available – data lost
101   No sample/reading taken – equipment out of action/unable to visit equipment
102   Sample lost or inadvertently discarded
103   Partial loss of sample
104   Sample frozen when collected
105   Snow in funnel when sample collected
106   Snow during sampling period
107   Bird droppings in funnel
108   Insects in sample
109   Leaves in sample
110   Soil in sample
111   Unidentified debris in sample
112   Debris in funnel
113   Sample discoloured
114   Bonfire in vicinity during sampling period
115   Heather burning in vicinity during sampling period
116   Forest fire in vicinity during sampling period
117   Straw burning in vicinity during sampling period
118   Crop spraying in vicinity during sampling period
119   Construction work in vicinity during sampling period
120   Liming in vicinity during sampling period
121   Change of land use in vicinity
122   Funnel assembly replaced with clean one
123   Clip open
124   Tubing damaged
125   Connections loose
126   River frozen – no sample
127   River dry – no sample
128   Fauna in tube
129   Condensation in tube
130   Plot/transect section not surveyed
131   Trampling during sampling period
132   Plot not accurately relocated
133   Evidence of disease in plot/transect section
134   Significant disturbance in plot/transect section
135   Pitfall trap flooded
136   Cell not surveyed due to obstruction
137   Flooding of survey area
138   Mowing of survey area
139   Muck/slurry/slag application
140   Application of chemicals (details should be supplied)
141   Grazing by sheep
142   Grazing by cattle
143   Grazing/browsing by deer
144   Grazing – other

145	Woodland management: coppicing
146	Woodland management: thinning
147	Woodland management: clearfelling
148	Woodland management: brushing
149	Windthrow
150	Animal path present
151	Forest ride present
152	Human path present
153	Path present (unspecified)
154	Rolling of survey area
160	Pitfall trap damaged
161	Pitfall trap lost
162	No antifreeze left in pitfall trap
163	Trampling by cattle during sampling period
164	Pitfall trap contained enough debris/mud to affect catch
165	Pitfall trap raised above ground level
166	Small amount of debris/mud in pitfall trap may have obscured small ground predators
167	Pitfall trap lid removed during sampling period
168	Trampling by sheep during sampling period
169	High river flow following snowmelt
170	Pond frozen
171	Spawn possibly killed by frost
172	Number of spawn masses too large to count
173	Number of spawn masses approximate
174	Pond dried up
175	No frogs seen prior to spawning
176	Pond covered in snow
177	No newly metamorphosed frogs seen leaving pond
178	Spawn stranded and dried out due to falling water levels
179	Pond drying up – only small surface puddles remain
180	Trace rainfall recorded (marked as 'X' on Met Office form)
181	Water in funnel, through no rain fell ('XX' on Met Office form)
182	Snow or sleet during sampling period
183	Wet bulb reservoir/wick frozen
184	Wet bulb reservoir/wick dry
185	Anemometer cups frozen
186	Rainfall catch likely to include some snow
187	Snow cleared from raingauge
188	Raingauge not level
189	Raingauge funnel blocked
190	Trace of snow recorded ('X' for snow depth on Met Office form)
200	Adverse weather conditions affected sampling/recording
201	Biting insects affected sampling/recording
202	Failing light affected sampling/recording
203	No flow observed in river – standing water only
204	Material inadequately preserved
205	Supplementary samples taken
206	Unidentified material archived
207	River stage iced up

208	Water sampling site cleared of weed/algae
210	Too few adult spittle bugs found for colour morph analysis
212	Small mammal(s) in trap – species unrecorded
213	Short-tailed field vole(s) in trap
214	Pygmy shrew(s) in trap
215	Frog(s) in trap
216	Common shrew(s) in trap
217	Common lizard(s) in trap
222	Non-standard sampling date
223	Non-standard sampling time
501	Laboratory: no sample
502	Laboratory: sample lost or inadvertently discarded
503	Laboratory: partial loss of sample
504	Laboratory: sample discarded because of contamination
505	Laboratory: insufficient sample for measurement
506	Laboratory: measurement not made because of equipment failure
507	Laboratory: sample pre-filtered
508	Laboratory: significant deposit of black material on filter
509	Laboratory: significant deposit of brown material on filter
510	Laboratory: significant deposit of green material on filter
511	Laboratory: no separate acidified sub-sample for Al and Fe

APPENDIX IV Key to abbreviations for measurement units

Abbreviation	Units
m	metre
cm	centimetre
mm	millimetre
μm	micrometre
km	kilometre
l	litre
ml	millilitre
g	gramme
kg	kilogramme
ha	hectare
h	hour
min	minute
s	second
yr	year
GMT	Greenwich Mean Time
BST	British Summer Time
°C	degrees Celsius
m^2	square metre
m^3	cubic metre
$m\ s^{-1}$	metres per second
$m^3\ s^{-1}$	cubic metres per second ('cumecs')
$W\ m^{-2}$	Watts per square metre
$\mu S\ cm^{-1}$	micro-Siemens per centimetre
$mg\ l^{-1}$	milligrammes per litre
$mmol_c\ kg^{-1}$	millimole charge per kilogramme
$kg\ m^{-3}$	kilogrammes per cubic metre
$kg\ ha^{-1}$	kilogrammes per hectare
$mg\ kg^{-1}$	milligrammes per kilogramme

Chapter 4   THE ECN DATABASE

Introduction

Databases form the core information resource for long-term monitoring programmes like ECN, and should aim to provide a complete representation of all information gathered over their duration. This is becoming even more important because rapid developments in new technology for remote access mean that the use and interpretation of information rely less on personal contact with data providers. A meta-information system is an essential part of any such database: it is not sufficient simply to provide the data values themselves, but necessary to include their description, derivation, measurement parameters, and quality criteria. Long-term environmental research databases must be reliable and stable, secure over a long timespan, accessible but with access controls, and flexible enough to allow for spatio-temporal analyses of a range of variables.

Users of data on environmental change range from scientists to the general public. The ECN database is seen primarily as a long-term information resource for scientific research into the processes of environmental change. However, it also has an important role in providing UK Government departments and agencies with more immediate information about long-term trends and early warning of environmental extremes which may influence policy or require immediate action.

The ECN Central Co-ordination Unit (CCU) is responsible for data handling in ECN and for the management and development of the database. Figure 14 gives an overview of the ECN data management system; its main components are:
- data input: data capture at ECN sites, transfer by electronic mail (e-mail) and validation.
- database, meta-database and geographical information system (GIS)
- remote data access systems

1.   Data handling and database design

1.1 Data capture

ECN data capture requires mainly manual methods, recording on to standard field forms or alternatively on to maps where the measurement is concerned with spatial patterns over the site. Automatic methods are currently used to record hourly meteorological measurements and 15-minute river stage and discharge. Wherever possible and appropriate, existing data capture techniques and common coding schemes have been adopted to maintain ECN's comparability with external networks.

The use of global positioning systems (GPS) has been investigated for recording spatial information such as vegetation plot locations and

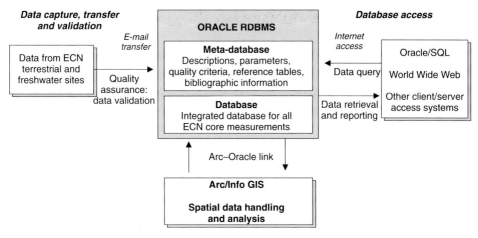

Data capture, transfer and validation

ORACLE RDBMS

Database access

Data from ECN terrestrial and freshwater sites

E-mail transfer

Quality assurance: data validation

Meta-database
Descriptions, parameters, quality criteria, reference tables, bibliographic information

Database
Integrated database for all ECN core measurements

Internet access

Data query

Data retrieval and reporting

Oracle/SQL

World Wide Web

Other client/server access systems

Arc–Oracle link

Arc/Info GIS

Spatial data handling and analysis

Figure 14. An overview of the ECN data management system

boundaries; a 'watching brief' is being kept over the reliability and cost-effectiveness of these systems for ECN. Recent developments in robust field computers for both form-based and map-based data entry are also being monitored for use when resources allow. Aerial photography and remotely sensed imagery (satellite or air) are important sources of information about changing spatial patterns over time both within ECN sites and in a wider context. All ECN terrestrial sites were flown in 1994 to give colour air photography at 1:10 000 scale; these will be digitised and incorporated within the ECN GIS. Options for using existing and also specially commissioned remotely sensed imagery are being explored.

ECN sites cover not only semi-natural environments but also intensively managed agricultural systems. Background data about the management of sites, particularly stocking rates, chemical applications, pest control and game management, are acquired where available as an important adjunct to the ECN measurements.

1.2 Data transfer

Site Managers are responsible for sending data to the CCU in machine-readable form using prescribed formats. Chapter 3 explains how sites, measurements and sampling events are identified, and specifies the sampling results, their units and precisions for each core measurement. Datasets are sent quarterly for frequent measurements taken throughout the year (15-minute to two-weekly), and six-monthly or yearly for seasonal measurements. E-mail is the preferred medium for transfer, with IBM format diskettes as an alternative where sites have no Internet connection. The ECN dataset format is comma-separated data values transferred as ASCII files – a format which the vast majority of software can handle and which is easily transferred by e-mail. Standard data entry forms with built-in validation procedures are being developed for manually recorded data. In addition to the measurement values, the specifications for most datasets include pre-defined 'quality codes' (listed in Chapter 3, Appendix III) which describe factors affecting measurement on a particular sampling occasion, or affecting a particular instrument or sample. In addition, sites may send free-text information where existing quality codes are insufficient or inappropriate. The generic ECN format and method for transfer, including how to handle missing data and associated quality information, is provided in Chapter 3, Section 2.

1.3 Data processing

Datasets are stored by the CCU in the form received and are logged by site, core measurement code and date, prior to processing. Receipt of e-mail datasets is acknowledged immediately; disk datasets generally await database input/validation, unless there is more than about two weeks' delay between receipt and further processing. Data loading, transformation and validation programs have been written for each dataset type. The time taken to process datasets is strongly affected by the type and number of errors they contain, and may vary between ten minutes and two days. Where possible, queries and problems are solved through communicating over e-mail, although, if errors are numerous or particularly difficult to decipher, then a request is made for a repeat dataset to be sent. Sites are strongly encouraged to check their data thoroughly before despatch; form-based data entry will undoubtedly help to minimise typing and formatting errors. Once processed and installed within the database, sites are sent an acknowledgement form which confirms input and reports any validation problems. Data validation methods are discussed separately under Section 2, **Quality assurance.**

ECN aims to have no more than a six-month time-lag between data collection and availability in the database, for measurements sampled throughout the year. Seasonal measurements should be available in the database by the end of June in the following year. The delay reflects resource levels available for data entry and processing, but more crucially it reflects the degree of quality assurance necessary to establish a long-term environmental research database. However, where a rapid response is required, instant access to recent data is provided with qualifications about the degree of validation performed and consequent data reliability.

1.4   Database system and design

A central database with remote network access was considered the most appropriate model for ECN to ensure a fully integrated system with the required data quality standards. The relational database management system (RDBMS) Oracle, with Structured Query Language (SQL) commands and utilities, forms the core software, with links to the GIS Arc/Info for spatial data handling. The database and software currently run on a Unix-based local area network (LAN) with high-speed links to the Internet for remote access and incoming data. The system provides an integrated storage and retrieval facility for all ECN data and associated meta-information.

The design of any database must focus not only on the data and their interrelationships but on the purpose and requirements of the activity it is to support. Where the styles of data are diverse and the range of uses difficult to predict, the system needs to be integrated, but as versatile as possible to:

- allow new structures to be generated and new datasets to be incorporated when required, and
- allow users free but guided access to data.

The range of potential uses of ECN data makes user requirements difficult to define and predict. Broadly, two main types of use are anticipated:

- scientific research, requiring access to high-resolution data, and the ability and freedom to define complex queries and analytical functions in space and time;
- information browsing and retrieval, requiring guided access to ECN information and summary data for display, extraction and incorporation into reports.

ECN ideally requires a fully integrated database, GIS and statistical system, which is also sufficiently open to allow links to external software. Whilst new software products exist which have begun to break down the boundaries between database and GIS functions and which use new and more versatile data models, these are still not widely adopted and their reliability and maintenance levels are uncertain. User requirements made it important that only tried and tested software formed the basis of the ECN database. However, new software developments will continue to be reviewed and incorporated where appropriate to maintain an acceptable balance between reliability and versatility.

Oracle provides the necessary access controls and allows the definition of suitable integrity rules. Arc/Info is well established as an analytical GIS for all types of spatial data, though GIS security and access controls tend to be less well developed than those of mainstream database software. The two systems can be linked – Oracle data can be viewed and analysed through Arc/Info – but are not integrated under the same management software. Oracle holds the majority of the ECN data, including grid references for point-based data, for easy access. Digital definitions of more complex spatial features, eg vegetation boundaries, field systems and river networks, are managed within Arc/Info. Attribute data for these features tend to be stored within Oracle and accessed *via* the Arc–Oracle link. Time-series digital maps will be established within Arc/Info to enable temporal analysis of changing spatial patterns.

Because apparent small changes in data capture method can radically alter a data structure, the database design for ECN had to be as straightforward and as flexible as possible. This is a continuing requirement, as the ECN programme is itself monitored for its effectiveness; changes in emphasis may be given to different types of measurement (whilst preserving long-term continuity as far as possible) which will have implications for the database structure. The core database stores raw data at the resolutions specified in the ECN Protocols. An associated summary database consists of monthly and/or annual summaries of the raw data, generated at six-monthly intervals. The database is logically divided into data and meta-data tables. Data for each replicate sampling location within an ECN site for a given core measurement are regarded as a logical 'dataset' which is allocated a unique identifier. A central meta-data table stores information about each dataset. Associated meta-data tables hold linked information on units of measurement, quality criteria, quality codes and text relating to sampling occasions, and reference tables for coded fields, eg species codes. The spatial and temporal dimensions of variables are important and affect the database structure: some measurements relate to an instance in time, eg surface water samples, whilst others are summary or cumulative values over a time period, eg rainfall or pitfall trap samples. Spatial data may relate to points, lines or areas, and in each case the location and spatial form need to be defined within the GIS.

Database security is an important consideration, in terms of both system faults and unauthorised access. Incremental back-ups of the database are made daily, a full back-up is made weekly, and monthly back-ups are kept for one year, off-site. Storage media are renewed regularly. Additional fail-safe devices, which maintain the status of the system and level of service for users in the event of disk crashes, are being considered. Access controls and security monitoring software are in operation to prevent unauthorised use.

2. Quality assurance in ECN

2.1 Introduction and terminology

Data of poor or unknown quality are unreliable. It is important to set quality standards at the outset of a data gathering exercise, but it is equally important to monitor how far those standards are met, and to ensure that this information accompanies the data for future use.

A number of different terminologies are commonly used to express different aspects of quality. For ECN purposes, the following terms seem the most appropriate, and are related to those used in the British and International Standards (British Standards Institution 1987,1995) for describing 'product' quality.

Quality assurance	A term embracing all planned and systematic activities concerned with the attainment of quality
Quality objectives	The specification of target values for quality criteria
Quality control	The practical means of ensuring that quality objectives are met, as set out in the specification

216

Quality assessment	(or quality verification). Procedures for assessment of the degree to which quality objectives have been met – ideally providing *evidence* that they have been met – after data capture. This should be a system of monitoring, which feeds back into the quality control process to maintain quality targets.

2.2 Quality control and assessment

The ECN Protocols are standard operating procedures (SOPs) designed to ensure consistency in measurement methods across sites and over time. They incorporate target specifications for quality criteria such as accuracy and recording resolution, where appropriate. Specifications for other quality criteria such as completeness and logical consistency of datasets (Association for Geographic Information 1996) are implicit within the Protocols and the formats for data transfer (available from the CCU). Quality control procedures go hand-in-hand with SOPs and have been included in the Protocols, eg correct handling of equipment and samples, maintenance schedules and calibration specifications, and unambiguous instructions for measurement and data handling. ECN has organised training courses to ensure that recorders are familiar with the documented procedures. During the first year of ECN monitoring in 1993, these courses helped to reveal some unforeseen practical difficulties in the Protocols which were amended accordingly.

Quality assessment should be regarded as a monitoring exercise to keep measurements 'on course' by feeding information back to the data capture stage. Where the measured feature can be kept, eg invertebrate samples, or re-visited, eg vegetation plots, the accuracy of identification may be assessed at a later date through subsampling by an independent 'expert'. Currently, ECN either uses known experts to identify all invertebrate specimens centrally, or sends them subsamples for verification. Randomly selected vegetation plots across the network are re-visited by an independent botanist shortly after survey.

The quality of more ephemeral measurements such as meteorological records can only be similarly assessed by running duplicate or parallel systems. Duplicate systems are expensive, and in practice assessment normally involves regular checks for instrument drift and recorder error. ECN runs manual daily weather stations (weekly at the less accessible sites) concurrently with automatic stations to provide some parallel records, and has regular maintenance schedules for equipment which help to maintain standards across the network and over time.

2.3 Data validation

Data validation can be regarded as part of quality assessment and involves screening data for 'unacceptable values' which may have occurred at any stage during data capture and handling. Present ECN data validation software for incoming data performs numeric range checks, categorical checks, formatting and logical integrity checks, eg on dates, number of samples, and links between datasets. Appropriate range settings for ECN variables have been selected following discussion with specialists in each field.

ECN adopts a cautious approach to discarding data on the principle that apparent errors may be valid outliers. Data values identified by validation software as 'unacceptable' are treated in one of three ways.

- Where values are clearly meaningless due to a known cause, eg an instrumentation fault, and cannot be back-corrected, the data are discarded and database fields set to null (no data).
- Where values are clearly in error, or out of range due to known calibration errors, and can be back-corrected, data are stored separately until the correction can be made.
- Where there is no straightforward explanation for outliers, the data are stored in the database, accompanied by meta-data 'flags' and associated text.

Sites are strongly encouraged to check their data before sending them to the CCU, but not to alter it unless a reason for error is clear, and in any case to inform the CCU of their actions.

The data validation checks described above are important 'first-pass' procedures, but they are relatively coarse and may fail to identify erroneous data within the valid range. An extreme example is a faulty instrument which generates values within the given range, but which have only two distinct values. More subtle problems may only be revealed through multivariate or time-series checks, based on known processes or expected patterns in the data. Procedures for implementing these as a 'second-pass' validation stage are currently being developed.

2.4 Laboratory practice

ECN sites send water samples to their own associated laboratories for analysis. The cost of standardising methods of analysis across all ECN laboratories is prohibitive; instead, analytical guidelines (see AG Protocol) list reference and approved techniques for each determinand, with corresponding limits of detection. Each laboratory practises its own internal quality control, and most participate in national quality assurance schemes. In addition, ECN has conducted inter-laboratory trials using standard solutions, and now operates a system whereby a standard solution sample accompanies each batch of water samples from the field. Analysis of the spread of values across the laboratories has helped to highlight problem areas; early attention to these areas has improved considerably the agreement between results.

2.5 Handling quality information in the meta-database

Target specifications for quality criteria are stored as meta-information alongside instrument and sampling details, and units of measurement. Any deviations from these specifications or from the sampling methods given in the Protocols are recorded with time-stamps. Details of laboratory methods and associated detection limits are stored similarly. Missing data and outliers which have been revealed by the quality assessment exercise but which cannot be corrected (see 2.3 above) are qualified, using pre-defined quality codes (listed in Appendix III, page 208) or free-text descriptions. This information may be associated either with a particular sampling occasion, or with an individual measurement variable on a sampling occasion. Site managers also use these codes or free text to describe factors affecting sampling outside their control, instrument damage or site management effects. Results of quality assessment exercises, eg laboratory trials or vegetation re-survey, are also incorporated. All meta-information may be linked directly with the data to which it relates.

3. Database access

3.1 Methods of access

In the past, large centralised databases have had a reputation for being relatively inaccessible and difficult to use. ECN's need to maintain close interaction between monitoring and scientific research means that database access is fundamental to the success of the programme. Although the ECN system has been centralised for reasons of integration and quality assurance, rapid developments in networking software over the past decade mean that this model no longer has the poor access implications of old, and the physical location of the database is less important. The challenge is to provide access modes to suit different styles of use and which can provide sufficient guidance and information about the system to enable little or no initial learning process for the user.

General-purpose database query and retrieval methods are provided primarily for scientific users already familiar with SQL and with the ECN database structures. Users may access the database *via* 'Telnet' and use SQL directly, or use a PC client front-end which constructs the SQL from a windows-style interface. However, this mode of access is unsuitable for users who require easy, guided access to the data without having to undergo prior training.

With the second type of user in mind, ECN has developed a 'tailored' interface to the ECN summary database using the World-Wide-Web (WWW). The WWW is an information service on the Internet which enables text and images, presented as Web 'pages', to be linked to other information through active areas of text, called 'hyperlinks'. The user can browse information in a more intuitive way, rather than being constrained to follow a pre-defined hierarchy of options. The system requirements are relatively low-cost: the user needs only a PC, an Internet link and 'browser' software to interpret the information and view the images. The addition of database links can generate a powerful information interface, enabling text, images and data to be presented together and allowing the user to progress from browsing information to guided data retrieval in a single system. ECN's Web pages (http://www.nmw.ac.uk/ecn/) incorporate a hyperlink to a database query, display and retrieval system (Brocklebank *et al.* 1996). This enables authorised users to build their own database query by selecting any combination of ECN sites, core measurement variables and date ranges for instant generation of tables and graphs. Data may also be downloaded *via* e-mail in 'column' format, for input to local software. Although the WWW software for presenting text and images is now very simple to operate, links to databases are less advanced so that significant programming effort was required to build the ECN user interface. However, generic software tools are developing rapidly and will greatly simplify the linking of databases and presentation of data. The ECN Web interface will continue to be enhanced to take advantage of new features and to provide new facilities.

ECN also provides access to 'real-time' data *via* the Web from an automatic weather station (AWS) at the Moor House–Upper Teesdale site in the north Pennines. Data are transmitted hourly *via* a modem link to the CCU and automatically displayed as graphs and tabulations. Direct links from AWSs at other ECN sites are planned, as well as from other automatic monitoring instrumentation, eg water quality loggers. Once these links are considered reliable enough, they will be used to download data automatically into the database through the data validation and input software.

3.2   Data ownership and access agreements

ECN data are owned jointly by the ECN sponsoring organisations which have agreed a system of user licensing and authorisation for access to high-resolution and summary data. Applications for access should be sent to the CCU, from where they are sent to the sponsoring organisations for consideration. If access is permitted, users are asked to sign a licence agreement which defines the terms under which ECN data may be used.

A.M.J. Lane
ECN Database Manager

References Association for Geographic Information. 1996. *Guidelines for geographic information content and quality.* London: AGI.

British Standards Institution. 1987. *Quality assurance.* (BSI Handbook 22.) London: BSI.

British Standards Institution. 1995. *Quality management and quality assurance: vocabulary.* (BS EN ISO 8402.) London: BSI.

Brocklebank, M., Lane, A., Watkins, J. & Adams, S. 1996. *Access to the Environmental Change Network summary database via the World-Wide-Web.* (ECN Technical Report 96/1.) Grange-over-Sands: Environmental Change Network.

Printed in the United Kingdom for The Stationery Office

Dd 302973 12/96 C7 G559 10170